Engineering Materials

This series provides topical information on innovative, structural and functional materials and composites with applications in optical, electrical, mechanical, civil, aeronautical, medical, bio- and nano-engineering. The individual volumes are complete, comprehensive monographs covering the structure, properties, manufacturing process and applications of these materials. This multidisciplinary series is devoted to professionals, students and all those interested in the latest developments in the Materials Science field, that look for a carefully selected collection of high quality review articles on their respective field of expertise.

More information about this series at http://www.springer.com/series/4288

Arash Ahmadivand · Burak Gerislioglu ·
Zeinab Ramezani

Toroidal Metamaterials

Fundamentals, Devices, and Applications

Springer

Arash Ahmadivand
Department of Electrical
and Computer Engineering
Rice University
Houston, TX, USA

Burak Gerislioglu
Department of Physics and Astronomy
Rice University
Houston, TX, USA

Zeinab Ramezani
Department of Electrical
and Computer Engineering
Northeastern University
Boston, MA, USA

ISSN 1612-1317 ISSN 1868-1212 (electronic)
Engineering Materials
ISBN 978-3-030-58287-6 ISBN 978-3-030-58288-3 (eBook)
https://doi.org/10.1007/978-3-030-58288-3

This Springer imprint is published by the registered company Springer Nature Switzerland AG
The registered company address is: Gewerbestrasse 11, 6330 Cham, Switzerland

Preface

The twenty-first century belongs to photonics. Many important industries from chip manufacturing and lighting, health care, and life sciences to astronomy and security rely on the same fundamental mastery of light. Future technologies will push for a steep increase in subwavelength photonic integration and energy efficiency, far surpassing that of bulk optical components, silicon photonics, and plasmonic circuits. Such a level of integration can be attained by embedding the data processing and waveguiding functionalities at the level of material rather than a chip, and the only conceivable solution to address those challenges is to utilize the recently emerged concepts of *metamaterials* and *metadevices* based on structuring of the matter at the subwavelength scales. In practice, metamaterials have enabled important capabilities ranging from subwavelength focusing to the ability of controlling magnetic response of nonmagnetic materials. Driven by the ongoing race to miniaturization, researchers are now able to devise ultracompact and integrated devices using such unique artificial architectures. In this context, one of the intriguing phenomena is the resonant scattering when synchronicity of resonant transmission and reflection can be condensed to the interference of discrete resonant levels with a continuum of nonresonant propagation modes. The exquisite result of this feature is the formation of asymmetric resonant lineshapes with narrow linewidth and low mode volumes. Among them, well-engineered toroidal metastructures have revitalized the performance of modern plasmonic and photonic tools through robust electromagnetic field confinement down to extreme subwavelength scales. Toroidal resonant systems are well known for generating unconventional gyrotropic-fashioned charge-current excitations' fingerprints with drastically concealed far-field radiation patterns. Toroidal moments provide physically substantial nonzero contributions to the fundamental properties of matter and scattered radiation. With their unconventional physical properties beyond classical multipoles, toroidal resonances possess a broad range of applications including but not limited to immunobiosensing, telecommunication, and nonlinear meta-optics.

This book enables readers to obtain the comprehensive understanding of toroidal metastructures and metadevices concepts. The book has six chapters in total encompassing diverse theoretical and applied aspects of the toroidal resonances' manifestation. The chapters were written by the experts from USA, who have extensively contributed to the evolution of science and engineering of the toroidal resonances in optical and terahertz metasystems. By considering the physics behind the formation of these unique multipoles, it is shown that toroidal resonances exhibit novel phenomena both in the linear and nonlinear response of photonic and plasmonic metamaterials that can significantly intensify the electromagnetic field confinement in both metallic and all-dielectric platforms, enabling the development of next-generation all-optical and optoelectronic instruments. By introducing the toroidal excitations in matter as an independent family of multipoles and evaluating their properties in comparison to classical electric and magnetic multipoles, the required theoretical framework of this phenomenon is presented. Besides, the multipolar expansion of toroidal multipoles is comprehensively discussed. The radiation efficiency, angular momentum loss, and recoil force of the toroidal multipoles are theoretically defined and described through the physical mechanism behind the toroidal multipoles.

In the book, the reader can find the principles of the excitation of toroidal resonances in both 3D metamaterials and quasi-infinite metasurfaces. It is demonstrated that these well-engineered architectures can be tailored in such a way to efficiently support pronounced toroidal modes at diverse frequency ranges. In light of this, there is a particular chapter that discusses the studies of the excitation of toroidal dipole modes in all-dielectric and plasmonic metamaterials and metasurfaces. Employing the defined insights in this book, one can understand the practical applications and technical instrumentation of toroidal metaplatforms in detail and the development of various devices based on this concept, such as nonlinear lasers, immunosensors, and photodetectors. For the nonlinear harmonic signal generation, taking advantage of ultratight confinement of electromagnetic field in the toroidal meta-atoms, the generation of deep ultraviolet light is presented. Similarly, the use of significant localization of plasmons and ultranarrow lineshapes of toroidal resonances in the advancement of ultraprecise metasensors with exceptionally improved limit of detection is introduced. Since, at the position of toroidal resonance, the scattering cross section of classical multipoles suppresses and the absorption spectra enhances. By considering these effects, the immense potential of toroidal metasurfaces in designing high photon yield light sensing tools is explained. Furthermore, the possibility of inducing strong coupling between the plasmonic modes in toroidal metasurfaces and the excitonic levels from quantum emitters is presented by explaining the mechanism behind this.

We hope that the book will be a valuable aid to comprehend the current research of the toroidal resonance phenomena in optical and terahertz structures for scientists, researchers, and graduate students working in the fields of optical engineering, applied physics, electrical engineering, materials science, and condense matter physics. We would like to express our appreciation to Dr. Anthony Doyle, Executive Editor, Springer, for his initial support of the book proposal and collaboration with us during preparation of the book. We are thankful to Laura Burgess, Katrin Petermann, and Boopalan Renu from the Springer Production Department for their assistance at the book production.

Houston, TX, USA Arash Ahmadivand
Houston, TX, USA Burak Gerislioglu
Boston, MA, USA Zeinab Ramezani

Contents

1 Introduction and Overview 1
1.1 The History of Light and Matter 1
1.2 The History of Light-Matter Interactions 3
1.3 The Discovery and Properties of Artificial Media 4

2 Classical Electrodynamics 7
2.1 Fundamental Principles of Static Electromagnetics 7
2.1.1 Coulomb's and Gauss's Laws 7
2.1.2 Biot-Savart and Ampère's Laws 9
2.1.3 The Lorentz Force 12
2.2 Equations for Static Fields 12
2.3 Fundamental Principles of Dynamic Electromagnetics 13
2.3.1 Maxwell's Equations in Vacuum 13
2.3.2 Maxwell's Equations in Macroscopic Media 14
2.4 The Electric Dipole 14
2.4.1 Multipole Expansion and Electric Multipoles 16
2.4.2 The Dipole and Quadrupole Potentials 18
2.5 Magnetic Multipoles 19
2.6 Unconventional Multipoles 23
2.6.1 Static Toroidal Multipoles 24
2.6.2 Dynamic Toroidal Multipoles 26
2.6.3 Dynamic Anapoles 35
References 37

3 Expansion of Electromagnetic Multipoles 41
3.1 Debye Potentials 41
3.2 Electromagnetic Radiations of Toroidal Solenoids 44
3.2.1 The Multipole Decomposition 44
3.2.2 The Dynamic Toroidal Multipoles 48
3.2.3 The Long Wavelength Regime 50

3.2.4 The Magnetostatic Regime 52
3.2.5 The Point Source Regime 53
References 54

4 Physical Mechanism Behind the Toroidal Multipoles 55
4.1 Defining the Problem 55
4.1.1 Potentials and Fields of a General Source 58
4.1.2 Fields at Far Distances 70
4.2 Radiation Intensity 81
4.3 Angular Momentum Loss 84
4.4 Recoil Force 95
4.5 The Connection Between Cartesian and Spherical Components of the First Multipoles 104
References 107

5 Toroidal Excitations in Metamaterials 109
5.1 Toroidal Excitations in 3D Artificial Media 109
5.2 Toroidal Multipoles in Planar Artificial Media 115
References 120

6 Toroidal Metadevices 123
6.1 Photodetection: Enhancing the Responsivity Performance 123
6.2 Nonlinear Lasing: Deep Ultraviolet Source 127
6.3 Immunosensors: Beyond Conventional Detection Limits 131
6.4 Plexciton Dynamics: Intensifying Ultrastrong Coupling 134
References 137

Chapter 1
Introduction and Overview

Abstract Pioneering efforts in understanding the *light-matter interactions* at subwavelength scales date back to nineteenth and twentieth centuries with the demonstration of electromagnetic wave propagation by breaking the diffraction limit and enhancing near-field effects for light localization. Studies showed that judicious interaction between light and matter leads to the emergence of resonant properties, which encompass an immense domain of physical insights from classical to advanced quantum electrodynamics. While light and matter are different entities, they possess significant influence on each other through some sort of intermediary doer. This has previously been demonstrated by Albert Einstein's well-known equation, $E = mc^2$, in which both photon energy and matter are the main indicators of the identical entity that are related to each other by the square of the speed of light in vacuum. However, the inherent characteristics of light and matter reveal the difference between these entities, which make the interaction between them consequential. In this limit, the wavelength (λ) of light and geometry of the matter have the key role in defining the properties of these interactions.

1.1 The History of Light and Matter

Historically, light is an electromagnetic radiation within the band of the electromagnetic spectrum that can be perceived by the human eye. At the beginning of the eleventh century, Hasan Ibn al-Haytham articulated that a ray of light takes the easier and faster path while passing through a specific medium. This cornerstone notion is known as *Fermat's principle* or *the principle of least time*, as the link between ray optics and wave optics. Later, at the beginning of the seventeenth century, by applying this principle, Fermat proved that the ratio of the sines of the incident and refracted angles is constant, which is known as *Snell's law* today. During these years, some other significant findings were also reported by Francesco Grimaldi to understand the form of light beyond geometrical shadow boundary, which is defined through the pseudo rectilinear motion of light-particle flux. This phenomenon was also experienced by Robert Hooke who anticipated that light forms due to rapid vibratory motion of the very small particles of which regular matter is composed.

A. Ahmadivand et al., *Toroidal Metamaterials*, Engineering Materials,
https://doi.org/10.1007/978-3-030-58288-3_1

In addition, Hooke was strongly believed to the fact that light propagates outwards from the center of each tiny vibrating center in circular patterns, and these beam rays were trajectories at right angles to these circular patterns. Although there had been a major progress in comprehending the characteristics of light until the onset of the eighteenth century, there were still considerable uncertainties about the nature of light. At that time, some of the fundamental questions about this notion were partially responded by Thomas Young through conducting diffraction measurements. This set of experiments considered that light waves possess vibratory motion along the longitudinal direction. In spite of this promising progress, the polarization of light was still unclear, and this obscurity was addressed by Augustin-Jean Fresnel. He proved that the polarization is consistent with the wave pattern when the periodic vibration become transverse to the propagation direction. In the meantime, Michael Faraday was conducting a set of measurements to define magnetic influences acting on bodies that are not directly in contact as lines of forces. These lines of forces were the starting point of force fields acting on bodies through space with the absence of any physical contact.

In the middle of nineteenth century, all these electric, magnetic, and optical phenomena were combined by James Clark Maxwell, and published in a book entitled as *A Dynamical Theory of the Electromagnetic Field*. The main scope of this context was proposing a set of equations with the assumption that light was a transverse electromagnetic wave, where these equations and the *Lorentz force* law constitute a unified classical theory of electricity, magnetism, and light. All of these achievements in the field of electromagnetism and light, as well as the brilliant discoveries in chemistry by John Dalton and Dimitri Mendeleev together led to the establishment of a clear distinction between light and matter. In that respect, Einstein proposed the quantization of radiation that reconstituted some of previously defined Newtonian corpuscular characteristics of light. His proposal showed that energy quanta in the form of $E = h\nu$, where ν is the frequency of light and h is the Planck's constant, removed the *ultraviolet catastrophe* from the *Rayleigh-Jeans formulation of black-body radiation.*

Ultimately, in the late nineteenth and first decades of the twentieth centuries, with the rise of *quantum-mechanics principles* by Max Born, Erwin Schrödinger, Werner Heisenberg, and Paul Dirac, the periodic table of elements was explained in terms of the inner structure of atoms. It was clearly indicated that quantum mechanics, in its non-relativistic form, is a reliable approach for perceiving matter outside the atomic nucleus. When combined with Dirac's initial formulation (i.e., quantized version of electromagnetism), the second revolution in physics was performed. Later, in 1916, Einstein successfully generalized his theory of relativity and applied it to gravity, which was a new geometric description of space–time for astrophysical energy and length scales.

1.2 The History of Light-Matter Interactions

Besides the advances in understanding of the characteristics of light and matter, the interaction between these entities was theoretically studied at the beginning of twentieth century. Gustav Mie was one of the first physicists who worked on the calculation of scattering of an electromagnetic wave from a homogenous dielectric sphere. He employed Maxwell's electromagnetic theory and applied it to spherical particles to theoretically understand the excitation of exceptional resonances, known as *surface plasmon resonances*, in gold colloids. Technically, *Mie scattering* for particles much smaller than the wavelength of the incident light proved that the major part of light-matter interaction is centered on the metallic components, and the excitation of coherently oscillating electrons at the interface between conductors and dielectrics is the direct result of such an interference. At the same time with Gustav Mie, Peter Debye also investigated the influence of the incident light intensity on the interaction with small particles. Although Mie theoretically demonstrated the excitation of resonances in metallic objects, the concept of surface plasmon resonances was not defined until the beginning of the last decade of twentieth century. Thomas Ebbesen was the pioneer scientist who experimentally validated that when light passes through a thin gold layer with subwavelength holes, the intensity of light can be boosted inside the holes. In addition to this advancement, in 1998, Peter Wolf showed that the oscillating electrons at the metallic surface can create a two dimensional sea of waves, or surface plasmons. He also proved that the excitation of these resonances can only be occurred when the energy and momentum of incoming photons and surface plasmons are equivalent. The results reported by Wolf verified that for a given metallic surface, which contains semi-infinite number of holes, the energy and momentum of plasmons can be manipulated in such a way that to interact with the visible light photons.

As a promising counterpart of modern photonics, plasmonics has received copious interest and witnessed a rapid progress in implementing next-generation all-optical and optoelectronics technologies. While the promise of plasmonics and the subsequent concepts have been well-acknowledged both theoretically and experimentally, only a small fraction of their prospective applications has reached to the commercial phase. This is due to the internal damping (e.g., resistive heating or Ohmic losses) and far-field radiation characteristics of metallic particles. Nevertheless, it is difficult to overestimate the impact of plasmonics on the optical society, therefore, several strategies have been conducted to control and reduce the lossy behavior of nanoscale metallic objects at high-energy bands, such as amplifying near-field coupling to diminish the far-field radiation. Indeed, the inevitable thermal heating in metallic structures has been addressed and employed efficiently in several research areas, such as photothermal therapy, cancer therapy, nanosurgery, and nanobubble formation. Beyond that, the discovery of optothermally controllable (e.g., phase-change materials) and optoelectronically tunable (e.g., two dimensional monolayers) materials and the combination of them with plasmonic architectures led to substantial advancements in the performance and functionality of photonic devices.

The need for much more efficient, low-loss, high-performance, and miniaturized optical instruments is accompanied with the development of breakthrough insights by merging diverse interdisciplinary techniques and principles. One of the leading motivations for this purpose is to use loss-less all-dielectric resonant particles and carefully engineered structures. For instance, the use of high-refractive index all-dielectric subwavelength structures in the implementation of modern technologies has been offered as an alternative approach instead of the lossy and metallic optical tools.

1.3 The Discovery and Properties of Artificial Media

Ongoing investigations in both classical and modern optical physics have led to the intriguing findings, which revitalized the performance of developed instruments by enabling manipulation and confinement of electromagnetic radiations down to extreme-subwavelength scales. These features were realized with the rise of artificial materials that possess properties beyond conventional substances in nature. In 1967, based on theoretical understandings, Victor Veselago demonstrated the possibility of designing a material that enables simultaneous negative values for both electric permittivity ($\varepsilon < 0$) and magnetic permeability ($\mu < 0$). Following, in 1987, Eli Yablonovitch verified the control and manipulation of light in artificial crystals. However, most of these structured photonic objects were designed to operate at wavelengths on the order of the lattice parameter until the development of *metamaterials*. By the discovery of metamaterials (after World War II), it is shown that this concept can be characterized as artificial media composed of periodically arranged subwavelength unit cells in order to attain the required electromagnetic functionality. Sir John Pendry (in 1990s) discovered that the radiation characteristics did not stem from the molecular or chemical structure of a given material. Instead, this property originates from the physical shape of the structures. He recognized rather than traditionally altering a material through its chemistry, the behavior of a material can be redefined by changing its physical properties. Negative index of refraction, optical magnetism, asymmetric transmission, hyperbolic dispersion, epsilon near-zero (ENZ), topological states, arbitrary control of light's trajectories, and cloaking are some of the prominent examples that can be realized by metamaterials across a wide spectral range from vacuum ultraviolet (VUV) wavelengths to terahertz (THz) frequencies. In addition, from the theoretical optical physics perspective, these manufactured systems with conventional resonant responses have been described using *multipole expansion framework*, in which the electromagnetic media can be signified by a series of dot electric and magnetic multipolar sources.

In general, the Lorentzian formula has been utilized to describe the spectral properties of metamaterials, where the induced lineshape can be considered as the summation of the intensities of individual resonances that contribute to the original lineshape. Indeed, the conventional model for the resonant behavior of metamaterials does not include the influence of interfering resonances. In 1961, this deficiency was addressed by the emergence of *Fano resonances*, by Ugo Fano, in quantum mechanical studies, and later it has been employed to encompass and exhibit the interference between resonances in optical systems. From the perspective of optical physics, the interference between different energy levels of a quantum–mechanical system is a global effect, analogous to the constructive and destructive interferences between classical waves. In principle, such quantum interferences occur in three-level atomic systems and lead to the excitation of sharp spectral features. Electromagnetically induced transparency (EIT) is an appropriate example for this concept, which can be distinguished as a sharp transmission window. This optical effect in metasurfaces significantly changes the dispersive properties of an opaque medium that can give rise to ultra-slow light phenomena and enhanced nonlinear effects at single photon power levels. Besides the classical Mie-type resonances, taking the advantage of artificial media, modern electrodynamics have witnessed the rise of other types of resonant modes with radiative and nonradiative properties, such as plasmonically induced transparency (PIT), charge transfer plasmons (CTPs), toroidal multipoles, etc. All of these advancements in photonics have underpinned important technologies from telecommunications to information processing and spectroscopy. Analysis of such radiations is an essential feature to investigate the properties of various substances, in which the scattered electromagnetic fields can be utilized as a reliable source to achieve noteworthy information for a given material. This requires for an accurate expansion of the radiated complex electric and magnetic multipoles. In condensed-matter, optical, atomic, nuclear, and solid-state physics studies, this phenomenon has been studied under the multipole decomposition framework. However, when we consider the electric and magnetic multipolar radiations, they should be investigated considering much more complicated patterns composed of oscillating charges and loop currents, respectively. As an alternative approach to the conventional multipole representations, the optically driven (dynamic) toroidal multipoles were introduced as a third and independent family of elementary electromagnetic sources. In 1957, Y. B. Zeldovich established the toroidal topology and studied this notion in several contexts from nuclear to molecular physics and classical electrodynamics, as well as solid state physics. Besides, dynamic nonradiating charge-current arrangements and nonreciprocal interactions have also been considered as other theoretical frameworks to describe the toroidal electrodynamics in the field of electromagnetism. In this Book, we introduce, demonstrate, and analyze the unique theoretical properties of toroidal multipoles and their applications in practical and real-world nanophotonic technologies.

Chapter 2
Classical Electrodynamics

Abstract Classical electrodynamics primarily deals with electromagnetic fields and their interactions caused by macroscopic distributions of charges and currents. In a specific mathematical process, this implies that the charge and current distributions can be confined in infinitesimally small volumes of space. In this Chapter, we firstly introduce the static electric and magnetic fields, and demonstrate how the conservation of electric charge and its relation to electric current leads to the dynamic connection between electricity and magnetism. Then, we discuss how these two were judiciously combined under classical electrodynamics, described by couple of dynamic field equations—the Maxwell equations.

2.1 Fundamental Principles of Static Electromagnetics

In this section, we briefly summarize the underlying and important principles in classical electrodynamics. This includes Coulomb's, Gauss's, Ampère's, and Biot-Savart laws, as well as introduction to Maxwell equations.

2.1.1 Coulomb's and Gauss's Laws

In experimental physics, the Coulomb's law (Plimpton and Lawton 1936; Jackson 1998; Greiner 2012; Franklin 2017) corresponds to the quantification of the amount of force between two stationary, electrically charged particles (Fig. 2.1). The electric force between these two charged objects is known as *Coulomb force* or *electrostatic force*, which can be described in a vectorial format as:

$$\mathbf{F}_{12} = \frac{Q_1 Q_2}{r^2}\mathbf{e}_r \tag{2.1}$$

where Q_1 and Q_2 are the magnitudes of the charges, r is the distance between two bodies, and $\mathbf{e}_r$ is the unit vector in the direction from Q_2 to Q_1. In a scalar format,

A. Ahmadivand et al., *Toroidal Metamaterials*, Engineering Materials,
https://doi.org/10.1007/978-3-030-58288-3_2

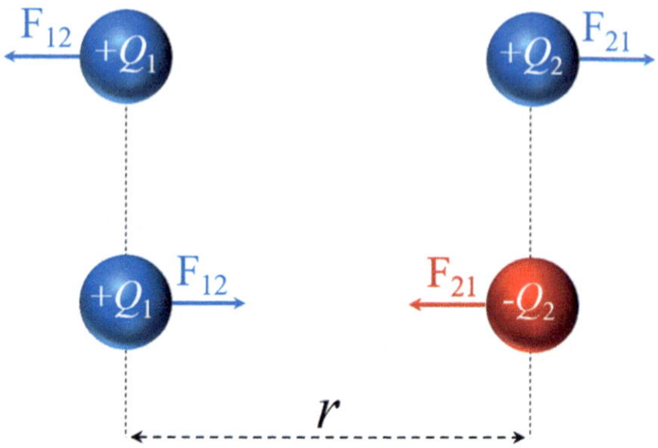

Fig. 2.1 Coulomb's law: charges with similar signs possess a force that pushes them further apart and charges with opposite signs possess a force that attracts them

$\mathbf{e}_r$ must be substituted by Coulomb constant ($k_e \approx 8.987 \times 10^9\ \mathrm{N \cdot m^2 \cdot C^{-2}}$). If the charges hold the same sign, the Coulomb force is repulsive ($F > 0$), and conversely, for the charges with opposite signs, the force is attractive ($F < 0$). The force on a given charge Q defines the electric field vector, as following:

$$\mathbf{F} = Q\mathbf{E} \tag{2.2}$$

Therefore, Coulomb's field law due to a source charge Q' is:

$$\mathbf{E} = \frac{Q'}{r^2}\mathbf{e}_r \tag{2.3}$$

Given that the electromagnetic field is linear, hence, the field due to a specific number of charges can be calculated by the summation of individual fields. However, in the presence of more than one source, the system becomes more complex and one must deal with two overlaid coordinate systems. This comprises the expressions for both the location of the charges and the location of the point where the field is being observed. A simple representation of this system is plotted in Fig. 2.2, in which $\mathbf{r}'$ and $\mathbf{r}$ locate the source point (Q_i) and the field point (p), respectively. Here, $\mathbf{r} - \mathbf{r}'$ defines the distance from a particular source to a field point, and the corresponding unit vector in this direction is $\mathbf{e}_{r-r'}$. Considering (2.3), the Coulomb's field law for an arbitrary array of point charges can be written in a general form as following:

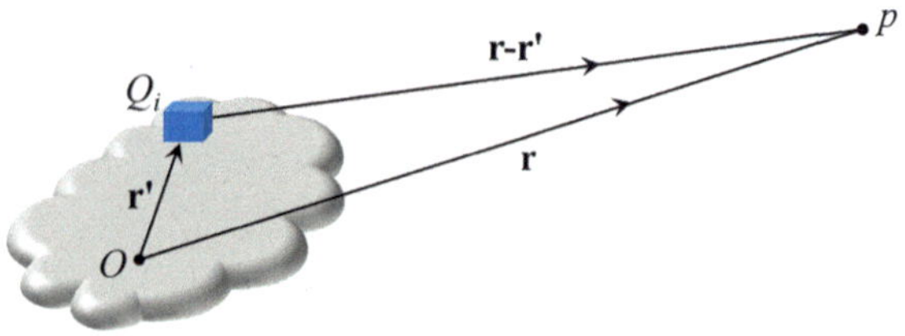

Fig. 2.2 The coordinate system for the source and field points

$$\mathbf{E}(\mathbf{r}) = \sum_i \frac{Q(\mathbf{r}'_i)}{|\mathbf{r} - \mathbf{r}'_i|^2} \mathbf{e}_{r-r'_i} \tag{2.4}$$

where i corresponds to the ith charge at location $\mathbf{r}'_i$. To define the volumetric description for Fig. 2.1 based on Coulomb's law, one needs to utilize the integration of the electric field (**E**) over the surface area by considering that the charge Q is enclosed by a Gaussian surface (S). The flux of **E** through this surface turns out to be equal to 4π times the enclosed charge. Thus, Gauss's law for **E** can be applied to explain the total net charge enclosed ($Q_{enclosed}$) within the surface as:

$$\oint_S \mathbf{E} \cdot d\mathbf{a} = 4\pi Q_{enclosed} \tag{2.5}$$

where the direction of $d\mathbf{a}$ is that of the outward normal to S. By transforming the integral form into a differential form for the surface with the volume of V, we have $Q_{enclosed} = \int_V \rho dv$, where ρdv is the three-dimensional charge-per-volume, and by substituting this into the following equation:

$$\oint_S \mathbf{E} \cdot d\mathbf{a} = \int_V \nabla \cdot \mathbf{E}\, dv = 4\pi \int_V \rho dv \tag{2.6}$$

we have the differential expression of Gauss's law as $\nabla \mathbf{E} = 4\pi \rho$.

Then (2.4) can be written as:

$$\mathbf{E}(\mathbf{r}) = \int_V \frac{\rho(\mathbf{r}')}{|\mathbf{r} - \mathbf{r}'|^2} \mathbf{e}_{\mathbf{r}-\mathbf{r}'} dv' \tag{2.7}$$

where ρ is the charge density, and the potential can be written as:

$$\Phi(\mathbf{r}) = \int_V \frac{\rho(\mathbf{r}')}{|\mathbf{r} - \mathbf{r}'|} dv' \tag{2.8}$$

2.1.2 Biot-Savart and Ampère's Laws

The fundamental laws of magnetism are rigorously complex. In 1820, Ørsted was the first person who qualitatively proved that currents produce magnetic fields (Oersted 1820). Later, the team of Biot and Savart as well as Ampère conducted the quantitative experiments of this insight. This resulted in the rise of two basic formulations for

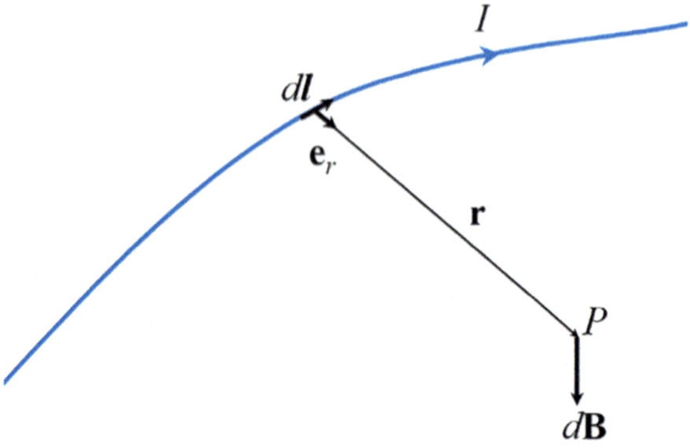

Fig. 2.3 The representation of Biot-Savart principle

steady currents, which is well-known as *Biot-Savart law*:

$$\mathbf{B} = \frac{1}{c} \oint \frac{\mathbf{e}_r \times I dl}{r^2} \tag{2.9}$$

and its integral form is known as *Ampère's law*, in which the path of integration overlaps with the physical current loop (Γ):

$$\oint_{\Gamma} \mathbf{B} \cdot dl = \frac{4\pi}{c} I_l \tag{2.10}$$

where I_l is the summation of all currents linking the Ampèrian loop and $\mathbf{e}_r$ is the unit vector that points from the source element (Idl) to the field point (P). Figure 2.3 demonstrates a schematic for the Biot-Savart law. Since steady-state currents flow in closed loops, there is an explicit topological difference between current loops that link the integration loop and those that do not.

Interestingly, Biot-Savart law is the magnetic analog of Coulomb's law, while Ampère's law is the magnetic analog of Gauss's law. Although the electric description of these laws is pretty simple, the magnetic case is much more complicated, due to the vectorial quantity of elementary magnetic sources.

In continue, by considering Ampère's law, the integral element can be described in differential form through the use of Stokes' theorem for an open surface (S) that is bounded by the curve (Γ):

$$\int_S (\nabla \times \mathbf{B})\, d\mathbf{a} = \oint_{\Gamma} \mathbf{B} \cdot dl \tag{2.11}$$

By assuming $\int_S \mathbf{J} d\mathbf{a} = I$, the Ampère law can be written as:

$$\int_S (\nabla \times \mathbf{B})\, d\mathbf{a} = \frac{4\pi}{c} \int_S \mathbf{J} \cdot d\mathbf{a} \tag{2.12}$$

Here, the surface S is arbitrary and therefore, the differential expression of Ampère's law for the total current can be written as:

$$\nabla \times \mathbf{B} = \frac{4\pi}{c} \mathbf{J} \tag{2.13}$$

Next, the magnetic field (**B**) is expressed in terms of vector potential of the electromagnetic field (**A**). To this end, one can make a connection between Figs. 2.1 and 2.3 through substituting **r** by (**r** − **r′**), and calculate the gradient of (**r** − **r′**) with respect to the coordinates of the field point as:

$$\nabla\left(\frac{1}{|\mathbf{r} - \mathbf{r}'|}\right) = \frac{-e_{\mathbf{r}-\mathbf{r}'}}{|\mathbf{r} - \mathbf{r}'|^2} \tag{2.14}$$

Then, the Biot-Savart law can be defined as:

$$\mathbf{B}(\mathbf{r}) = -\frac{I}{c} \oint_{\Gamma'} \nabla\left(\frac{1}{|\mathbf{r} - \mathbf{r}'|}\right) \times d\mathbf{r}' \tag{2.15}$$

By removing gradient from the integral, the following terms can be extracted:

$$\begin{cases} \mathbf{B} = \nabla \times \left(\frac{1}{c} \oint \frac{Idl}{r}\right) \\ \mathbf{A} = \frac{1}{c} \oint \frac{Idl}{r} \end{cases} \tag{2.16}$$

Eventually, the above-mentioned equations lead to $\mathbf{B} = \nabla \times \mathbf{A}$. In fact, the definition between the magnetic field and vector potential refers to a unique relationship between the scalar potential and electric field. On the other hand, according to magnetic Gauss's law, the divergence of the curl of any vectorial function is zero, thus $\nabla \cdot \mathbf{B} = 0$. This principle clearly indicates that there are no isolated magnetic poles in nature, hence, magnetic field lines form closed loops which never begin or end. Although the lines of **B** are bounded curves in an ideal condition, these lines are not closed in actual situations.

So far, the obtained results were appropriate for a one-dimensional system consisting of current along a wire. However, the **B** and **A** would be more complicated for a two- (current-per-width on a surface) and three-dimensional (current-per-width in a volume) systems. In this limit, one can write:

$$\begin{cases} \mathbf{B} = \frac{1}{c} \int_V \frac{\mathbf{J}(\mathbf{r}') \times \mathbf{e}_{r-r'}}{|\mathbf{r}-\mathbf{r}'|^2} dv' \\ \mathbf{A} = \frac{1}{c} \int_V \frac{\mathbf{J}(\mathbf{r}')}{|\mathbf{r}-\mathbf{r}'|} dv' \end{cases} \tag{2.17}$$

which will be utilized in the following subsections extensively. On the other hand, the macroscopic description of Ampère's law in a magnetic medium is an important concept, where free currents should be distinguished from bound magnetization currents. This notion will be applied in the subsequent sections.

2.1.3 The Lorentz Force

The magnetic force on a moving charge with the velocity of **u** can be:

$$F_m = \frac{q\mathbf{u}}{c} \times \mathbf{B} \tag{2.18}$$

Now, by applying an electric force to the moving charge, the total force can be obtained through the following equation:

$$F_t = q\left(\mathbf{E} + \frac{\mathbf{u}}{c} \times \mathbf{B}\right) \tag{2.19}$$

where F_t is known as *Lorentz force* on a moving charge and it is valid for the steady-state and time-varying electromagnetic fields.

2.2 Equations for Static Fields

Until now, we have summarized most of the fundamental equations for static fields in steady-state conditions. Below, one can find a list of well-defined macroscopic field equations for a free charge and current:

$$\nabla \cdot \mathbf{D} = 4\pi\rho \quad \text{(Gauss's law for electric fields)} \tag{2.20}$$

$$\nabla \times \mathbf{E} = 0 \quad \text{(Conservative nature of electrostatic fields)} \tag{2.21}$$

$$\nabla \cdot \mathbf{B} = 0 \quad \text{(Gauss's law for magnetic fields)} \tag{2.22}$$

$$\nabla \times \mathbf{H} = \frac{4\pi}{c}\mathbf{J} \quad \text{(Ampère's law)} \tag{2.23}$$

in which, **D** (as $\mathbf{D} = \varepsilon\mathbf{E}$) and **H** (as $\mathbf{H} = 1/\mu\mathbf{B}$) correspond to the fundamental fields **E** and **B**, respectively, and ρ and **J** are interpreted as total charge and current density for linear, isotropic media. In addition to the static fields equations, picturing the force on moving charges and explanation the set of equations would be useful to understand the Maxwell's equations. Since all current loops must have a closed-loop form in steady-state conditions, one can write $\nabla \cdot \mathbf{J} = 0$. Besides, due to Ohm's law for conductive materials, the following relation between electric field and current density can be established: $\mathbf{J} = \sigma \mathbf{E}$.

2.3 Fundamental Principles of Dynamic Electromagnetics

Thus far, we outlined some of the basic equations and mathematical description of static electric and magnetic fields. However, these explanations are not reliable if the field quantities are time-dependent, where the electric and magnetic fields cannot be separated and discussed individually. Therefore, a generalized concept is required to address this limitation. As mentioned at the beginning, James Clerk Maxwell utilized the results of Faraday's studies to develop time-dependent electromagnetic field equations. Obtained from these experimental studies, the proposed set of equations by Maxwell are the mathematical abstractions that describe a broad range of phenomena and clearly represent the classical electromagnetic field. Here, in light of previously defined laws in static electromagnetics including scalar and vector potentials of the fields, we will briefly explain dynamic electromagnetics using Maxwell's equations.

2.3.1 *Maxwell's Equations in Vacuum*

The time-dependent Maxwell's equations for electromagnetic fields with free charges and densities are listed below:

$$\nabla \cdot \mathbf{D} = 4\pi\rho \tag{2.24}$$

$$\nabla \cdot \mathbf{B} = 0 \tag{2.25}$$

$$\nabla \times \mathbf{E} + \frac{1}{c}\frac{\partial \mathbf{B}}{\partial t} = 0 \tag{2.26}$$

$$\nabla \times \mathbf{H} - \frac{1}{c}\frac{\partial \mathbf{D}}{\partial t} = \frac{4\pi}{c}\mathbf{J} \tag{2.27}$$

The above equations can also be written in the form of total current and density by substituting these components with total sources. Here, $\mathbf{B} = \mu_0\mathbf{H}$ and $\mathbf{D} = \varepsilon_0\mathbf{E}$ are the external sources in vacuum.

2.3.2 *Maxwell's Equations in Macroscopic Media*

Beyond time-dependent Maxwell's equations for electromagnetic fields and sources in vacuum, the equations for **E** and **B** can be thought as they are valid everywhere in space. This assumption is true for definite number of sources, because the observation of the fields is feasible. However, for macroscopic aggregates of matter, solving these equations is quite challenging. For averaged quantities of the macroscopic fields and sources, the macroscopic Maxwell equations have the form of the (2.20)–(2.24) with **E** and **B**, which are the averaged **E** and **B** of the microscopic or vacuum Maxwell equations. Conversely, **D** and **H** are no longer simple multiples of **E** and **B**, respectively, and the macroscopic field quantities **D** and **H** possess the following components:

$$D_\alpha = \varepsilon_0 E_\alpha + \left(P_\alpha - \sum_\beta \frac{\partial Q'_{\alpha\beta}}{\partial x_\beta} + \cdots \right) \tag{2.28}$$

$$H_\alpha = \frac{1}{\mu_0} B_\alpha - (M_\alpha + \cdots) \tag{2.29}$$

in which P, M, and $Q'_{\alpha\beta}$ represent the macroscopically averaged electric dipole, magnetic dipole, and electric quadrupole modes, respectively. Similarly, the charge and current densities are macroscopic averages of the free charge and current densities in the medium.

As mentioned above, in the macroscopic limit, the Maxwell's equations are based on **E**, **B**, **D**, and **H**. These homogenous equations can be solved by expressing **E** and **B** in terms of the scalar potential (**Φ**) and vector potential (**A**). Nevertheless, the inhomogeneous equations cannot be solved until deriving **D** and **H** in terms of **E** and **B**, which is well-acknowledged as **D** = **D**[**E**, **B**] and **H** = **H**[**E**, **B**].

In general, and in most of materials, the electric and magnetic quadrupolar and high-order terms are weak and negligible, therefore, the electric and magnetic polarization, **P** and **M**, respectively, correlating with the dipolar mode are significant and will be considered. However, this does not imply that the constitutive relations are simple.

2.4 The Electric Dipole

An electric dipole can be exemplified simply by considering a static system composed of two charges of equal magnitude with opposite signs, located with the distance of l from the origin. In Fig. 2.4, the potential at the point P (r, θ, φ) is given by:

$$\Phi(r, \theta, \varphi) = Q\left(\frac{1}{R_1} - \frac{1}{R_2}\right) \tag{2.30}$$

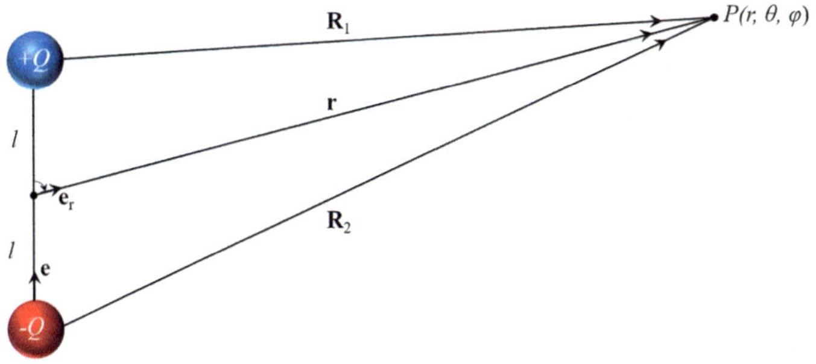

Fig. 2.4 An elementary system for a charge-pair dipole

Expressing the potential in terms of the magnitude of r ($|\mathbf{r}|$) and θ, assuming $l \ll r$, and using the cosine law, one can write the potential in the following format:

$$\Phi(r,\theta) = 2Ql\frac{\cos\theta}{r^2} \tag{2.31}$$

Importantly, because the charge-distribution is axially symmetric, the potential must be independent of the azimuthal angle (φ). As can be seen from (2.31), the potential due to a dipole decreases by a factor of $1/r^2$, and the potential due to a single charge reduces by a factor of $1/r$. This stems from the fact that by moving away from the observation point P, the dipole charge distribution appears as a small unit with zero charge. Moreover, the electric dipole moment of a pair of equal charges can be described by $\mathbf{p} \equiv 2Ql\mathbf{e}$, in which $\mathbf{p}$ is the dipole moment and as shown in Fig. 2.4, its direction is defined depending on the position of negative and positive charges. Now, by considering $\mathbf{e}_r$ as the unit vector in the direction of the field point and applying (2.31), the dipole potential can be expressed as:

$$\Phi = \frac{\mathbf{p}\cdot\mathbf{e}_r}{r^2} \tag{2.32}$$

and the electric field vector of the dipole can be defined as $\mathbf{E} = -\nabla\cdot\Phi$. Then, using the spherical components of the electric field vector as provided below, one can plot the dipole equipotential and field lines as demonstrated in Fig. 2.5.

$$\begin{aligned} E_r &= 2p\tfrac{\cos\theta}{r^3} \\ E_\theta &= p\tfrac{\sin\theta}{r^3} \\ E_\varphi &= 0 \end{aligned} \tag{2.33}$$

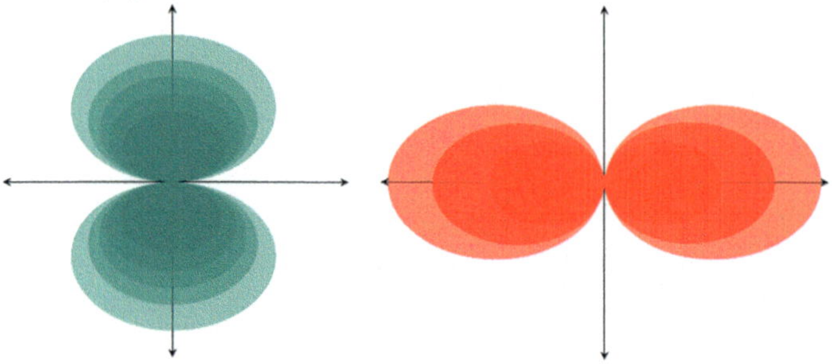

Fig. 2.5 Electric dipole equipotential and electric field lines

2.4.1 Multipole Expansion and Electric Multipoles

Multipole expansion is a reliable method to describe the potential due to an arbitrary distribution of charges. Here, we initially consider the expansion of multipoles and later apply this principle to understand the classical electric multipoles. By considering a static collection of arbitrary charges (Q_α) and defining the position of changes as $\mathbf{r}'_\alpha(x'_{\alpha,i})$ (see Fig. 2.6), one can specify the potential at the field point due to the source through the following relation: $\Phi_\alpha = Q_\alpha \big/ R_\alpha$. However, by simplifying the calculations and limiting ourselves to a fixed field point, the Taylor expansion with respect to the coordinates $\mathbf{r}'_\alpha = (x'_{\alpha,1}, x'_{\alpha,2}, x'_{\alpha,3}, x'_{\alpha,4})$ can be written as:

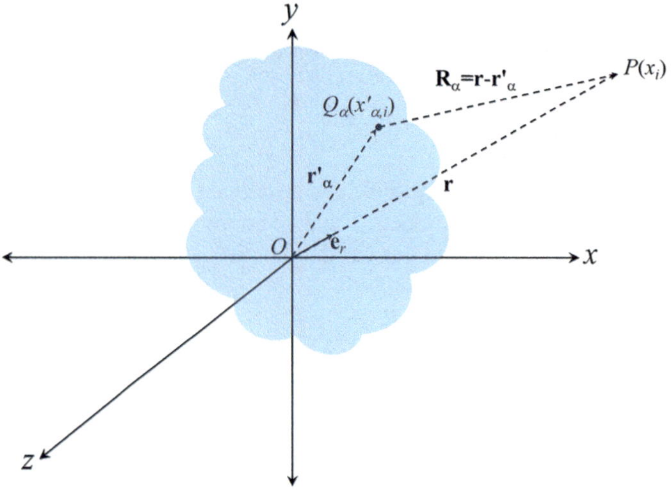

Fig. 2.6 A diagram showing the source and field points

$$f(\mathbf{r}'_\alpha) = C + \sum_i x'_{\alpha,i}\left[\frac{\partial f(\mathbf{r}'_\alpha)}{\partial x'_{\alpha,i}}\right]_{\mathbf{r}'_\alpha=0} + \frac{1}{2}\sum_{i,j} x'_{\alpha,i}x'_{\alpha,j}\left[\frac{\partial^2 f(\mathbf{r}'_\alpha)}{\partial x'_{\alpha,i}\partial x'_{\alpha,j}}\right]_{\mathbf{r}'_\alpha=0}$$
$$-\frac{1}{6}\sum_{i,j,k} x'_{\alpha,i}x'_{\alpha,j}x'_{\alpha,k}\left[\frac{\partial^3 f(\mathbf{r}'_\alpha)}{\partial x'_{\alpha,i}\partial x'_{\alpha,j}\partial x'_{\alpha,k}}\right]_{\mathbf{r}'_\alpha=0} + \cdots \tag{2.34}$$

where C is the constant for the first term $f(0)$. In continue, using $f(\mathbf{r}'_\alpha) = Q_\alpha / R_\alpha(\mathbf{r}'_\alpha)$, the equation above can be utilized to derive the potential due to the charge Q_α as:

$$\Phi_\alpha = \frac{Q_\alpha}{r} + Q_\alpha \sum_i x'_{\alpha,i}\left[\frac{\partial}{\partial x'_{\alpha,i}}\left(\frac{1}{R_\alpha}\right)\right]_{R_\alpha=r}$$
$$+\frac{1}{2}Q_\alpha \sum_{i,j} x'_{\alpha,i}x'_{\alpha,j}\left[\frac{\partial^2}{\partial x'_{\alpha,i}\partial x'_{\alpha,j}}\left(\frac{1}{R_\alpha}\right)\right]_{R_\alpha=r}$$
$$-\frac{1}{6}Q_\alpha \sum_{i,j,k} x'_{\alpha,i}x'_{\alpha,j}x'_{\alpha,k}\left[\frac{\partial^3}{\partial x'_{\alpha,i}\partial x'_{\alpha,j}\partial x'_{\alpha,k}}\left(\frac{1}{R_\alpha}\right)\right]_{R_\alpha=r} + \cdots \tag{2.35}$$

Consequently, the potential due to a collection of charges can be simply written as:

$$\Phi = \sum_\alpha \Phi_\alpha = \Phi^{(1)} + \Phi^{(2)} + \Phi^{(4)} + \cdots + \Phi^{(2^l)} \tag{2.36}$$

in which each potential component can be described as follows:

$$\Phi^{(1)} \equiv \sum_\alpha \frac{Q_\alpha}{r}$$
$$\Phi^{(2)} \equiv -\sum_\alpha q_\alpha \sum_i x'_{\alpha,i}\frac{\partial}{\partial x_i}\left(\frac{1}{r}\right)$$
$$\Phi^{(4)} \equiv \frac{1}{2}\sum_\alpha q_\alpha \sum_i x'_{\alpha,i}x'_{\alpha,j}\frac{\partial^2}{\partial x_i \partial x_j}\left(\frac{1}{r}\right)$$
$$\vdots$$
$$\Phi^{(2^l)} \equiv \frac{(-1)^l}{l!}\sum_\alpha q_\alpha \sum_{i,j,\ldots,l} x'_{\alpha,i}x'_{\alpha,j}\ldots x'_{\alpha,l}\frac{\partial^l}{\partial x_i \partial x_j \ldots \partial x_l}\left(\frac{1}{r}\right) \tag{2.37}$$

In (2.36) and (2.37), the first term ($\Phi^{(1)}$) corresponds to the monopole potential, and the second ($\Phi^{(2)}$) and third ($\Phi^{(4)}$) terms are related to the dipole and quadrupole potentials, respectively.

2.4.2 The Dipole and Quadrupole Potentials

For the dipole potential, one must consider the second term of (2.37) ($\Phi^{(2)}$) and rewrite the equation as below, where the first term represents the dipole moment:

$$\Phi^{(2)} = -\underbrace{\sum_{\alpha} q_{\alpha}\mathbf{r}'_{\alpha}}_{dipole\,moment} \cdot \nabla\left(\frac{1}{r}\right) \tag{2.38}$$

Therefore, the dipole potential can be simplified as $\Phi^{(2)} = \mathbf{p}\cdot\mathbf{e}_r/r^2$. Besides, by taking the gradient of the dipole potential, the electric dipole field vector can be specified as $\mathbf{E}^{(2)} = -\nabla\left(\Phi^{(2)}\right)$, which leads to:

$$\mathbf{E}^{(2)} = \frac{1}{r^3}\left[3(\mathbf{p}\cdot\hat{\mathbf{r}})\hat{\mathbf{r}} - \mathbf{p}\right] \tag{2.39}$$

In continue, we focus on the potential due to arbitrary distribution of charges for the third term in (2.37). Employing transform modification based on the inertia tensor in rigid body dynamics, one can assume that $1/r$ is the solution of Laplace's equation, except $r \neq 0$, and therefore, for $r > 0$:

$$\sum_i \frac{\partial^2}{\partial x_i^2}\left(\frac{1}{r}\right) = 0 \tag{2.40}$$

which can be written including the delta function:

$$\sum_{i,j} \frac{\partial^2}{\partial x_i \partial x_j}\left(\frac{1}{r}\right)\delta_{ij} = 0 \tag{2.41}$$

Hence, the equation for the quadrupole potential can be defined as:

$$\Phi^{(4)} = \frac{1}{6}\sum_{i,j} Q_{ij}\frac{\partial^2}{\partial x_i \partial x_j}\left(\frac{1}{r}\right) \tag{2.42}$$

where Q_{ij} is a 3×3 array matrix, called *quadrupole tensor*, that can be stated as: $Q_{ij} \equiv \sum_{\alpha} Q_{\alpha}\left(3x'_{\alpha,i}x'_{\alpha,j} - r'^2_{\alpha}\delta_{ij}\right)$. Given that the tensor matrix is symmetric, therefore, one can assume that $Q_{ij} = Q_{ji}$. To simplify the calculations, we consider the diagonal elements of the quadrupole tensor as: $Q_{kk} = \sum_{\alpha} Q_{\alpha}\left(3x'^2_{\alpha,k} - r'^2_{\alpha}\delta_{kk}\right)$. Summing over this reduced relation leads to:

$$\sum_k Q_{kk} = \sum_\alpha Q_\alpha \left(3 \left(\sum_k x'^2_{\alpha,k} \right) - r'^2_\alpha \left(\sum_k \delta_{kk} \right) \right) = 0 \tag{2.43}$$

where $\sum_k x'^2_{\alpha,k} = r'^2_\alpha$ and $Q_{kk} = \sum_k \delta_{kk} = 3$. Thus, the sum of diagonal elements of the matrix (i.e., trace) vanishes, and at most *five* components of the tensor matrix are independent. This results in much simpler theory and reduction of the independent elements to *two*. It is mathematically proved that the charge distribution possesses an axis of symmetry. Therefore, in this condition, we only have *one* independent element in the analysis. For instance, if we choose the x'_3 axis, then $Q_{11} = Q_{22}$, hence, the only remaining independent element would be Q_{33}. This quantity is often abbreviated as Q (i.e., the quadrupolar moment). Now, by assuming that the charge distribution is continuous rather than discrete, then we have:

$$Q = \int_V \rho(\mathbf{r}') \left(3x'^2_3 - r'^2 \right) dx'_1 dx'_2 dx'_3 \tag{2.44}$$

where $\rho(\mathbf{r}')$ is the charge density at the point defined by $\mathbf{r}'$ over the volume of V of the charge density distribution. Ultimately, the potential of a quadrupole can be formulated as:

$$\Phi^{(4)}(r, \theta) = \frac{1}{2} Q \frac{\left(\frac{3}{2} \cos^2 \theta - \frac{1}{2} \right)}{r^3} \tag{2.45}$$

where θ is the angle between the quadrupole's axis and observation point.

It is noteworthy to mention that in some cases, because of the specific geometry of a given charge distribution, certain multipole terms can be neglected. For instance, the monopole term disappears if there are equal amounts of positive and negative charge, and similarly, the dipole term vanishes whether the distribution composed of equivalent dipoles that are oppositely oriented. In Fig. 2.7, some of the important members of the electromagnetic multipoles family are illustrated. Here, the electric dipole, formed by two opposite charges separated by a distance l, is the simplest part of this configuration.

2.5 Magnetic Multipoles

In addition to the electric multipoles, magnetic multipoles constitute the remaining part of the electromagnetic multipoles, and in continue, we briefly discuss the multipole expansion of the magnetic effects of steady currents.

As a starting point, the vector potential can be defined as:

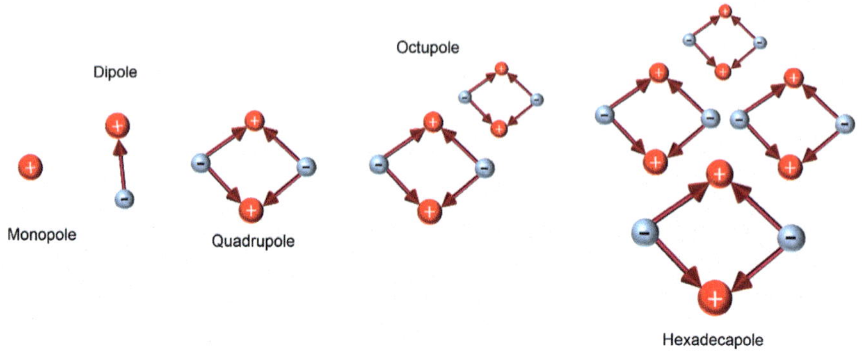

Fig. 2.7 A sequence of electric multipoles

$$A(r) = \frac{1}{c} \int\limits_{v} \frac{\mathbf{J}(\mathbf{r}')}{|\mathbf{r} - \mathbf{r}'|} dv \tag{2.46}$$

Here, by considering $|\mathbf{r} - \mathbf{r}'|$ as R, one can expand the term $1/R$ in the electrostatic limit, and the first two terms of (2.37) can be rewritten as following:

$$\mathbf{A}(\mathbf{r}) = \frac{1}{cr} \int\limits_{v} \mathbf{J}(\mathbf{r}')\, dv' - \frac{1}{c} \int\limits_{v} \mathbf{J}(\mathbf{r}') \left[\mathbf{r}' \cdot \nabla \left(\frac{1}{R} \right) \right] dv' + \cdots \tag{2.47}$$

Similar to what we have done in the electrostatic regime, one can formulate the monopole term as:

$$\mathbf{A}^{(1)} = \frac{1}{cr} \int\limits_{v} \mathbf{J}(\mathbf{r}')\, dv' \tag{2.48}$$

Under steady-state conditions, the current density in a given system can be deemed to rise from many closed filamentary current-loops. Thus, the integral above can be represented as the sum over the line integral of the filamentary current around each individual loop (Fig. 2.8):

$$\int\limits_{v} \mathbf{J}(\mathbf{r}')\, dv' = \sum_{\varsigma} \oint\limits_{\Gamma_\varsigma} I'_\varsigma\, d\mathbf{s}'_\varsigma \tag{2.49}$$

Here, for any current-loop, I'_ς is constant, hence, this component will be independent from the integral. Now (2.49) is:

$$\int\limits_{v} \mathbf{J}(\mathbf{r}') dv' = \sum_{\varsigma} I'_\varsigma \oint\limits_{\Gamma_\varsigma} d\mathbf{s}'_\varsigma \tag{2.50}$$

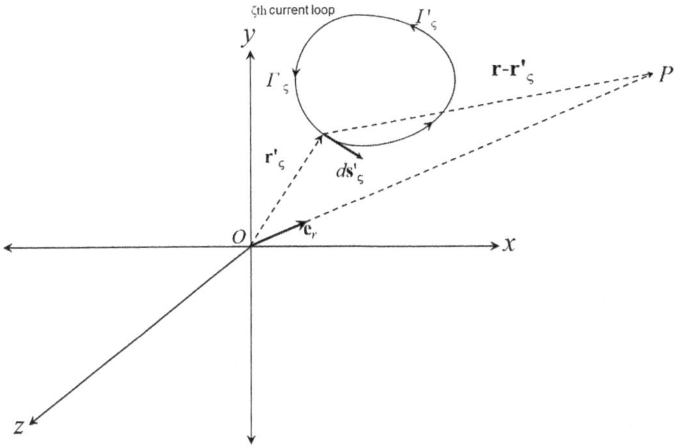

Fig. 2.8 A sample system indicating the current-loop sources

Given that the integrand is a differential, therefore, the right side of the equation vanishes. This fundamental outcome reveals that the magnetic multipole expansion does not contain a monopole term. Then, the second term in (2.47) can be transformed into:

$$\mathbf{A}^{(2)} = -\frac{1}{c}\sum_{\varsigma} I'_\varsigma \oint_{\Gamma_\varsigma} \mathbf{r}'_\varsigma \cdot \nabla\left(\frac{1}{r}\right) d\mathbf{s}'_\varsigma \tag{2.51}$$

Next, by assuming the above integral encompasses the projected area of the current-loop and by taking into account the bounded surface (S) in Fig. 2.9, one can claim that $\oint_\Gamma \mathbf{r}' \times d\mathbf{s} = 2\oint_S d\mathbf{a} = 2\mathbf{S}$, in which **S** is the vector that possesses the

Fig. 2.9 A representation of the current-loop area

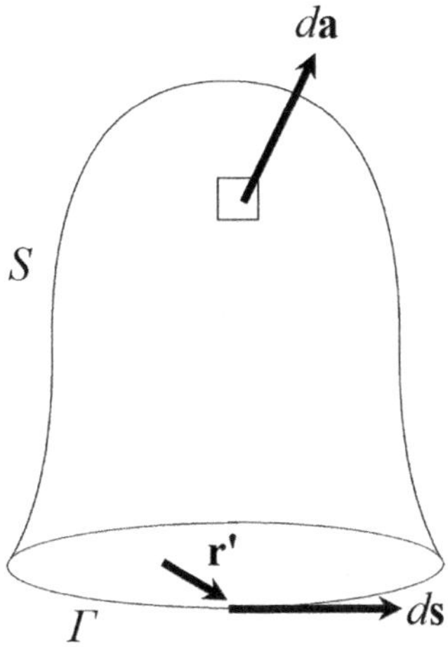

magnitude and orientation of the maximum projected area defined by the contour Γ. Mathematically, the cross-product of $(\mathbf{r}' \times d\mathbf{s})$ with an arbitrary vector $\mathbf{k}$ can be written as $(\mathbf{r}' \times d\mathbf{s}) \times \mathbf{k} = d\mathbf{s}(\mathbf{r}' \cdot \mathbf{k}) - \mathbf{r}'(d\mathbf{s} \cdot \mathbf{k})$, and since the integration step $(d\mathbf{s})$ along the current loop is equivalent to $d\mathbf{r}'$, therefore $d'(\mathbf{r}'(\mathbf{r}' \cdot \mathbf{k})) = d\mathbf{s}(\mathbf{r}' \cdot \mathbf{k}) + \mathbf{r}'(d\mathbf{s} \cdot \mathbf{k})$. Here, an integration over the closed-loop makes this equation is equivalent to zero, thus, the remaining terms must be equal or opposite to each other. Using this fact and taking into account (2.51), we have:

$$\mathbf{A}^{(2)} = -\frac{1}{c}\sum_{\varsigma} I'_{\varsigma} \oint_{\Gamma_{\varsigma}} \mathbf{r}'_{\varsigma} \cdot \nabla\left(\frac{1}{r}\right) d\mathbf{s}'_{\varsigma} = \frac{1}{2c}\sum_{\varsigma} I'_{\varsigma}\left[\nabla\left(\frac{1}{r}\right) \times \int \mathbf{r}' \times d\mathbf{s}'_{\varsigma}\right]$$

$$= \frac{1}{2c}\sum_{\varsigma} I'_{\varsigma}\left[\nabla\left(\frac{1}{r}\right) \times \int_{S_{\varsigma}} d\mathbf{a}'_{\varsigma}\right] \tag{2.52}$$

Ultimately:

$$\mathbf{A}^{(2)} = \left[\frac{1}{c}\sum_{\varsigma} I'_{\varsigma}\mathbf{S}_{\varsigma}\right] \times \left[\nabla\left(\frac{1}{r}\right)\right] \tag{2.53}$$

where $\mathbf{S}_{\varsigma}$ is the effective area of the ςth current-loop. To define the magnetic moment of current I' flowing through a plane that encloses an area S, we can write $\mathbf{m} = (I'/c)\mathbf{S}$. This implies that the first term in (2.53) is the sum of all of the elementary dipole moments $(\mathbf{m}_{\varsigma})$. Consequently, the (2.53) can be written in the form of:

$$\mathbf{A}^{(2)} = \sum_{\varsigma} \mathbf{m}_{\varsigma} \times \left[\nabla\left(\frac{1}{r}\right)\right] = \mathbf{m}_t \times \left[\nabla\left(\frac{1}{r}\right)\right] \tag{2.54}$$

Here, $\mathbf{m}_t$ is the magnetic dipole moment of the entire system of currents. By expanding the gradient in the equation above, one can write:

$$\mathbf{A}^{(2)} = \frac{\mathbf{m}_t \times \mathbf{e}_r}{r^2} \tag{2.55}$$

Given that there is no monopole term in the magnetic multipole expansion ($\mathbf{A}^{(1)} \equiv 0$), the lowest-order contribution to the magnetic field will be obtained by taking the curl of the dipole term in vector potential expansion, $\mathbf{B}^{(2)} \equiv \nabla \times \mathbf{A}^{(2)}$, which can be explained as following:

$$\mathbf{B}^{(2)} = \nabla \times \left(\frac{\mathbf{m}_t \times \mathbf{r}}{r^3}\right) \tag{2.56}$$

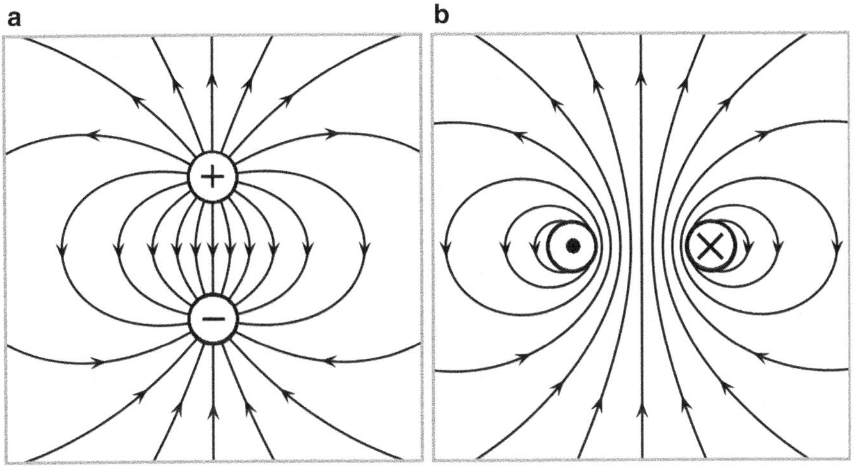

Fig. 2.10 **a** Electric and **b** magnetic dipoles

This equation can be expanded in the following form:

$$\mathbf{B}^{(2)} = \frac{1}{r^3}\nabla \times (\mathbf{m} \times \mathbf{r}) - (\mathbf{m} \times \mathbf{r}) \times \nabla\left(\frac{1}{r^3}\right) \tag{2.57}$$

which leads to:

$$\mathbf{B}^{(2)} = \frac{1}{r^3}\left[3(\mathbf{m} \cdot \hat{\mathbf{r}})\hat{\mathbf{r}} - \mathbf{m}\right] \tag{2.58}$$

Here, the magnetic dipole field possesses the same form as the electric dipole (2.39). As demonstrated in Fig. 2.10, in both cases, the internal fields are highly strong.

2.6 Unconventional Multipoles

So far, we have studied the properties of sources and fields, and a clear picture from the classical electric and magnetic multipoles is represented. The introduced framework for the multipole expansion has extensively been employed in several fields including but not limited to optical and condensed-matter phenomena. However, beyond point-like multipole sources, there is a third family of multipoles independent from the elementary electromagnetic sources, which are known as *toroidal* multipoles (Papasimakis et al. 2016). In 1957, toroidal moments were introduced by

Zel'dovich (1958) and have been applied and investigated in the context of nuclear, atomic, and molecular physics, classical electromagnetism, and solid-state physics (Dubovik and Cheshkov 1974; Flambaum and Khriplovich 1980; Dubovik and Tugushev 1990; Flambaum and Murray 1997; Ceulemans and Chibotaru 1998; Afanasiev and Dubovik 1998; Afanasiev 2001). In earlier studies, the concept of *toroidal electrodynamics* was established based on static toroidal moments, and recently, dynamic toroidal moments have emerged and explained under the classical electrodynamic framework. In the following subsections, we consider both static and dynamic toroidal multipoles by providing details on their formation mechanisms and exquisite properties.

2.6.1 Static Toroidal Multipoles

The static toroidal dipole, also known as *anapole* (i.e., without poles), was introduced for the first time in nuclear physics to explain parity-violating weak interactions. The result of this phenomenon was a radiationless arrangement of static currents flowing across the surface of a torus, equivalent to a ring of static magnetic dipoles aligned head-to-tail (see Fig. 2.11a). In theory, these currents create a magnetic field localized within the torus, thus static toroidal dipoles do not interfere directly with either electric or magnetic fields of the incidence, which is the fundamental difference between the anapole and classical static electric and magnetic moments. Generally, two types of static toroidal moments have been proposed in toroidal electrodynamics: electric

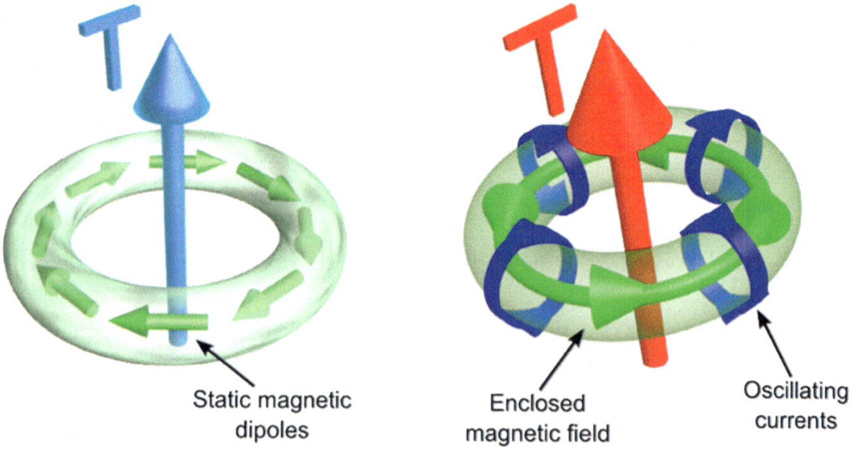

Fig. 2.11 **a** An artistic graph representing the static toroidal moment formed by a toroidal configuration of static currents and magnetic dipoles. **b** The dynamic toroidal dipole moment created by poloidal currents oscillating on the meridians of a torus, enclosing a confined magnetic field. Adapted from (Raybould 2017) with permission

(axial) and magnetic (polar). The axial moments stem from the vortex-like electric field configurations, whereas the polar moments originate from current configurations. Nevertheless, polar toroidal moments are merely considered, because of the absence of magnetic charge-currents in the established theoretical framework (Afanasiev 1990; Afanasiev and Stepanovsky 1995).

The anapole concept was successfully generalized to a family of toroidal multipoles and applied to the explanation of condensed matter by an order parameter, termed as *toroidization* or *toroidal polarization*, similar to the macroscopic electric polarization and magnetization in classical electromagnetism, where the electric polarization corresponds to the electric dipole density and magnetization corresponds to the magnetic dipole density. Indeed, toroidization specifies the density of toroidal dipoles and configurations of local toroidal moments with long-range orders are responsible for the formation of macroscopic toroidization. A medium with macroscopic toroidization is known as a *ferrotoroid*, equivalent to a ferroelectric and a ferromagnet. This was perceived for the first time during nuclear physics experiments of Cesium (Cs) atoms, through the precise measurements of the amplitude of parity-nonconserving transitions between 6*S* and 7*S* states of Cs with the use of a spin-polarized atomic beam (Wood et al. 1997). It was verified that the nucleus of Cs atoms supports a toroidal mode, comprising the currents flowing on the surface of a torus. In this limit, an electron in the *p*-orbital partially overlaps with the nucleus and would have interactions which break the *p*-symmetry (Haxton 1997). Particularly, such a toroidal moment in the nucleus has a partial contribution to parity violation in light interaction with the Cs atoms. The results for this set of experiments were consistent with the fact that ferroelectrics break the spatial inversion symmetry and ferromagnets break the time-reversal symmetry. Besides, ferrotoroids with magnetic toroidization break time-reversal and inversion symmetry, concurrently. Ferrotoroids demonstrate electric polarization in response to an external magnetic field, and conversely, demonstrate magnetization to external electric fields; with effective electric toroidization shaped by a loop of electric dipoles, invariant under time ($t \rightarrow -t$) and space ($r \rightarrow -r$) inversions (Naumov et al. 2004; Van Aken et al. 2007).

Static toroidal patterns have been observed in different systems based on diverse compounds including but not limited to metals, glasses, pyroxenes, boracites, olivines, ferroelectric nanoparticles, and molecular magnets (Toledano et al. 2011; Tokura 2007; Yamaguchi and Kimura 2013; Mettout et al. 2010; Hayami et al. 2014; Tolédano et al. 2015). The existence of toroidal currents and toroidal topology in various scales from atomic nuclei and organic molecules to extremely large astronomical scales (e.g., galaxies, neutron stars, quasars, black holes, etc.) is summarized in Fig. 2.12.

Beyond nuclear physics and condensed matter concepts, the formalism of toroidal electrodynamics was explained in solid state physics (Dubovik et al. 1986; Dubovik and Tugushev 1990). Toroidal ordering in this context was first discussed on ferromagnetic domains of small particles by Kittel in 1946 (Kittel 1946), and it was confirmed in real compounds known as boracites in 1974 (Zheludev et al. 1974). Pyroxenes and olivines were also verified as other compounds that sustain toroidal moments (Smith 1969; Hong-Jian and Fa-Min 2009).

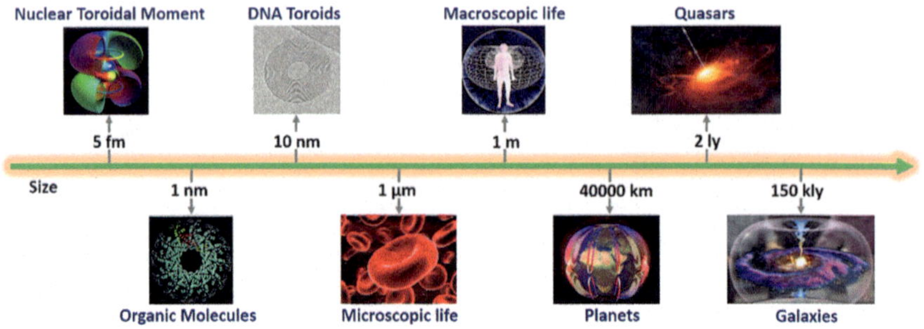

Fig. 2.12 Structures with different sizes from atomic to extremely large scales in physics, biology, chemistry, and astronomy that exhibit toroidal topology

2.6.2 *Dynamic Toroidal Multipoles*

After the observation of static toroidal modes in diverse systems, in the last decade of the 20th century, the discussion of dynamic toroidal multipoles has started with a deep focus on understanding the radiation properties of time-dependent toroidal current configurations. As explained in this Chapter, a fundamental feature in all types of toroidal moments' excitations is the *space-time symmetry*. In this limit, both electric dipoles ($\overrightarrow{p} = \sum_i Q_i\vec{r}_i$) and magnetic dipoles ($\vec{m} = \sum_i Q_i\vec{r}_i \times d\vec{r}_i/dt$) change their sign only with space inversion ($r \to -r$) and time reversal ($t \to -t$), respectively. Besides, one can simply find out that the radiative toroidal dipole moment as $\vec{T} = \sum_i Q_i\vec{r}_i \times (\vec{r}_i \times d\vec{r}_i/dt)$, which change its sign under either time reversal or spatial inversion (Radescu and Vlad 1998; Ögüt et al. 2012).

A famous and practical example that supports a toroidal moment is a solenoid, or similarly, a closed loop of azimuthally oriented magnetic moments. Such behavior of toroidal solenoids unveils the coupling between electric and magnetic dipoles. This (hybrid) nature of toroidal moments enables specific features to manipulate magnetic polarization by electric fields and vice versa. The theory behind the excitation of toroidal solenoids in different orders can be utilized to mathematically describe the radiation properties of time-dependent toroidal current arrangements as $\mathbf{j}(t, \mathrm{r}) = \nabla \times \nabla \times (c\mathbf{T}(t)\delta^{(3)}(\mathrm{r}))$ (Boardman et al. 2005). However, for simplicity, we initially limit ourselves to the space domain and try to extract the induced current due to toroidal moment with zeroth (0th) order (see Fig. 2.13a):

$$\overrightarrow{J} = \overrightarrow{n}_{\varphi} I\delta(\rho - d)\delta(z) \tag{2.59}$$

where d is radius of the loop and $\overrightarrow{n}_{\varphi}$ is a unit vector tangent to the plane of the loop. Since $\nabla \cdot \overrightarrow{J} = 0$, the current flowing for the zeroth mode can be imagined by an analogous magnetization concept ($\overrightarrow{M}$):

$$\overrightarrow{M} = \overrightarrow{n}_{z} I\zeta(d - \rho)\delta(z) \tag{2.60}$$

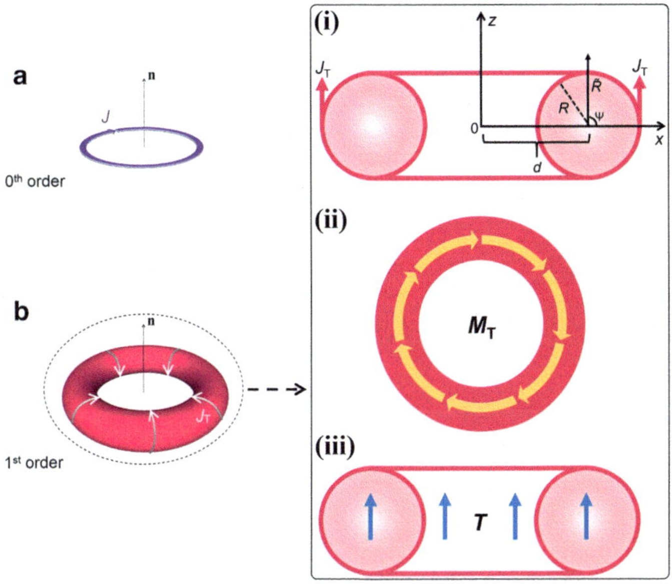

Fig. 2.13 **a** Current distribution representation of a circular current-loop for a magnetic dipole. **b** Current-carrying, firmly winded (where the helicity of the solenoid vanishes) toroidal solenoid of 1st order. It can also be considered as a source of a toroidal dipole. The arrows demonstrate the poloidal currents flowing on the surface of an imaginary torus (i). Here, the torus is intensively wrapped by the magnetized rings (ii). The directed toroidization along the axis of the torus (iii). Adapted from (Gerislioglu and Ahmadivand 2019) with permission. Copyright Multidisciplinary Digital Publishing Institute (MDPI)

where $\zeta(\square)$ is a step function and $\vec{J} = \nabla \times \vec{M}$. This equation can be understood as an expression of Ampère's theorem, in which the closed circular current is equal to the magnetic moment normal to it (Afanasiev 2001). Here, the induced magnetic field can be derived from (2.59) and (2.60), and the related magnetic vector potential can be written as:

$$\vec{B} = \frac{I}{c}\int \frac{1}{\left|\vec{r} - \vec{r}'\right|}\vec{n}_{\varphi}\delta(\rho' - d)\delta(z')dV' \tag{2.61}$$

For very small radii, $\vec{J}$ is not well-described, since the unit vector tangent is no longer valid at the origin, while $\vec{M}$ is still well-expressed. Besides, when d tends to zero ($d \to 0$) in (2.59), we will have a magnetic dipole moment as a result of very small values of d.

Considering a more complex system consisting of poloidal currents flowing on the surface of magenta-colored torus along its meridians (Fig. 2.13b), the induced current due to the 1st order mode can be written as (Afanasiev et al. 1996; Raybould 2017; Gerislioglu and Ahmadivand 2019):

$$\vec{J}_T = -\frac{\Phi c \delta\left(R - \tilde{R}\right)\vec{n}_\psi}{8\pi^2\left(d - \sqrt{d^2 - R^2}\right)\left(d + \tilde{R}\cos\psi\right)} \tag{2.62}$$

where $\vec{n}_\psi$ is the unit vector tangent to the magenta-colored torus surface ($\vec{n}_\psi = \vec{n}_z \cos\psi - \vec{n}_\rho \sin\psi$) and Φ is the magnetic flux which penetrates to the torus. For (2.62), the following conditions are assumed: $\tilde{R}$, ψ, and ϕ are the corresponding Cartesian coordinates:

$$\begin{cases} x = \left(d + \tilde{R}\cos\psi\right)\cos\phi \\ y = \left(d + \tilde{R}\cos\psi\right)\sin\phi \\ z = \tilde{R}\sin\psi \end{cases} \tag{2.63}$$

Moreover, the surface of the magenta-colored torus can be imagined where $R = \tilde{R}$. By varying ψ, ϕ, and for a fixed $\tilde{R}$, one can claim that the x, y, and z points mentioned above can fill the surface, where $(\rho - d)^2 + z^2 = R^2$ (Fig. 2.13b(i)). Analogous to the 0th mode case, for $\nabla \cdot \vec{J}_T = 0$, $\vec{J}_T$ (Fig. 2.13b) can be described in terms of magnetization ($\vec{M}_T$) (Fig. 2.13b(ii)), where $\vec{M}_T$ is confined within the magenta-colored torus and it has only ϕ component (which only changes inside the torus) as following:

$$\vec{M}_T = \frac{\Phi c \vec{n}_\phi}{8\pi^2 \rho\left(d - \sqrt{d^2 - R^2}\right)}\varsigma\left(R - \sqrt{(\rho - d)^2 + z^2}\right) \tag{2.64}$$

Since $\nabla \cdot \vec{M} = 0$, the resulting toroidization ($\vec{T}$) is: $\vec{M} = \nabla \times \vec{T}$, where $\nabla \cdot \vec{T} \neq 0$. Figure 2.13b(iii) illustrates the ensuing toroidization pattern. In light of the toroidization and (2.64), the poloidal current in (2.62) is equal to the magnetization and to the toroidization, which are the extensions of Ampère's law. Additionally, one can deduce the following as toroidization:

$$\begin{aligned} \frac{\vec{T}}{\vec{n}_z} = &\left(\frac{c\Phi}{8\pi^2\left(d - \sqrt{d^2 - R^2}\right)}\right) \\ &\left(\varsigma\left(d - \sqrt{R^2 - z^2} - \rho\right)\ln\frac{d + \sqrt{R^2 - z^2}}{d - \sqrt{R^2 - z^2}}\right. \\ &\left. + \varsigma\left(d + \sqrt{R^2 - z^2} - \rho\right)\varsigma\left(\rho - d + \sqrt{R^2 - z^2}\right) \times \ln\frac{d + \sqrt{R^2 - z^2}}{\rho}\right) \end{aligned} \tag{2.65}$$

Now, considering different regions in space, $\vec{T}/\vec{n}_z$ possesses different expressions:

$$\begin{cases} \dfrac{\overrightarrow{T}}{\overrightarrow{n}_z} = \dfrac{c\Phi}{8\pi^2\left(d-\sqrt{d^2-R^2}\right)} \ln \dfrac{d+\sqrt{R^2-z^2}}{d-\sqrt{R^2-z^2}}, \text{ where } 0 \le \rho \le \left(d - \sqrt{R^2 - z^2}\right) \\ \dfrac{\overrightarrow{T}}{\overrightarrow{n}_z} = \dfrac{c\Phi}{8\pi^2\left(d-\sqrt{d^2-R^2}\right)} \ln \dfrac{d+\sqrt{R^2-z^2}}{\rho}, \text{ where } \left(d - \sqrt{R^2 - z^2}\right) \le \rho \le \left(d + \sqrt{R^2 - z^2}\right) \end{cases} \tag{2.66}$$

Importantly, in other domains, $\vec{T}/\vec{n}_z$ is zero. By taking account of (2.65) and (2.66), one can conclude that extracting the toroidal dipole moment ($\overrightarrow{T}$) is quite challenging. To address this concern, a volume integral can be applied:

$$\int \overrightarrow{T} dV = \frac{c\pi R^2 d\Phi \overrightarrow{n}_z}{8\pi(d - \sqrt{d^2 - R^2})} \tag{2.67}$$

Until now, we have elucidated the theory behind the formation of zeroth and first order loop currents in space domain based on the toroidization principle. Similar approach can be applied to study the second and third order of loop currents due to toroidal excitations. However, to understand the influence of time into the explanation of toroidal electrodynamic framework, we summarize the general distribution of charge and current configurations using charge (ρ) and current (j) densities:

$$\frac{\partial \rho(\vec{r}, t)}{\partial t} + \vec{\nabla} \cdot \vec{j}(\vec{r}, t) = 0 \tag{2.68}$$

Parameterizing (2.68) in terms of three major components, including electric ($Q_{lm}(-k^2, t)$), magnetic ($M_{lm}(-k^2, t)$), and toroidal ($T_{lm}(-k^2, t)$) multipoles, allows to express ρ and $\vec{j}$ as a function of space and time as:

$$\rho(\vec{r}, t) = \frac{2}{8\pi^3} \sum_{l,m,k} (-ik)^l \frac{\sqrt{4\pi(2l+1)}}{(2l+1)!!} \times Q_{lm}\left(-k^2, t\right) F_{lmk}(\vec{r}) \tag{2.69}$$

$$\begin{aligned} \vec{j}(\vec{r}, t) = {} & \frac{2}{8\pi^3} \sum_{l,m,k} (-ik)^{l-1} \frac{\sqrt{4\pi(2l+1)(l+1)}}{\sqrt{l}(2l+1)!!} \\ & \times \left\{ kM_{lm}\left(-k^2, t\right) \vec{F}^{(0)}_{lmk}(\vec{r}) + \left[\frac{1}{c} \dot{Q}_{lm}(0, t) + k^2 T_{lm}\left(-k^2, t\right) \right] \right. \\ & \left. \times \vec{F}^{(+)}_{lmk} + \frac{1}{c} \sqrt{\frac{l}{l+1}} \dot{Q}_{lm}\left(-k^2, t\right) \vec{F}^{(-)}_{lmk}(\vec{r}) \right\} \end{aligned} \tag{2.70}$$

where

$$\sum_k = \int_0^\infty k^2 dk; \quad \{m = -l, \ldots, l; \quad l = 0, 1, 2, \ldots\} \tag{2.71}$$

In the equations above and in the upcoming relations, the sum over l starts from $l = 0$ for the electric multipole and $l = 1$ for the magnetic and toroid multipoles. The dot over Q_{lm} ($\dot{Q}_{lm}$) defines derivation with respect to t. Finally, the factor of $F_{lmk}(\vec{r})$ is the system of regular solutions of the Helmholtz equation (Gumerov and Duraiswami 2005):

$$\left(\Delta + k^2\right) F_{lmk}(\vec{r}) = 0 \tag{2.72}$$

$$F_{lmk}(\vec{r}) = j_l(kr) Y_{lm}(\vec{n}),\ \vec{n} = \frac{\vec{r}}{r} \tag{2.73}$$

$$j_l(kr) = \sqrt[3]{2\pi} i^l \frac{J_{l+1/2}(kr)}{\sqrt{kr}} \tag{2.74}$$

where j_l and $j_{l+1/2}$ are the spherical and cylindrical Bessel functions, respectively, and Y_{lm} are the classical spherical harmonics. Thus, the normalization and completeness conditions are:

$$\int F_{lmk}(\vec{r}) F *_{l'm'k'} (\vec{r}) d^3 r = \delta_{ll'} \delta_{mm'} \frac{8\pi^3}{k^2} \delta\left(k - k'\right) \tag{2.75}$$

$$\sum_{l,m,k} F_{lmk}(\vec{r}) F *_{l'm'k'} (\vec{r}) = 8\pi^3 \delta\left(\vec{r} - \vec{r}'\right) \tag{2.76}$$

$$\sum_k = \int_0^\infty k^2 dk \tag{2.77}$$

$$F_{lmk}(-\vec{r}) = (-1)^l F_{lmk}(\vec{r}) \tag{2.78}$$

and the basis factor functions $F_{lmk}(-\vec{r})$ are the solutions of vector Helmholtz equation as:

$$\left(\Delta + k^2\right) \vec{F}_{lmk}(\vec{r}) = 0 \tag{2.79}$$

$$\vec{F}^{(0)}_{lmk}(\vec{r}) = \frac{i}{\sqrt{l(l+1)}} \vec{\nabla} \times \left[\vec{r} F_{lmk}(\vec{r})\right] = j_l(kr) \vec{Y}_{llm}(\vec{n}) \tag{2.80}$$

$$\begin{aligned} \vec{F}^{(+)}_{lmk}(\vec{r}) &= \frac{-1}{\sqrt{l(l+1)}} \frac{i}{k} \vec{\nabla} \times \vec{\nabla} \times \left[\vec{r} F_{lmk}(\vec{r})\right] \\ &= \frac{1}{\sqrt{2l+1}} \left\{\sqrt{l}\, j_{l+1}(kr) \vec{Y}_{ll+1m}(\vec{n}) + \sqrt{l+1}\, j_{l-1}(kr) \vec{Y}_{ll-1m}(\vec{n})\right\} \end{aligned} \tag{2.81}$$

$$\vec{F}^{(-)}_{lmk}(\vec{r}) = -\frac{i}{k} \vec{\nabla} F_{lmk}(\vec{r})$$

$$= \frac{1}{\sqrt{2l+1}} \left\{ \sqrt{l}\, j_{l-1}(kr) \vec{Y}_{ll-1m}(\vec{n}) - \sqrt{l+1}\, j_{l+1}(kr) \vec{Y}_{ll+1m}(\vec{n}) \right\} \tag{2.82}$$

More detailed information on the corresponding spherical vectors can be found in Ref. (Radescu and Vlad 1998). Consequently, the electromagnetic and toroidal multipolar form factors can be defined as a function of previously defined equations:

$$Q_{lm}\left(-k^2, t\right) = \frac{(2l+1)!!}{(-ik)^l \sqrt{4\pi(2l+1)}} \int \rho(\vec{r}, t) j_l^*(kr) Y_l^*(\vec{n}) d^3r \tag{2.83}$$

$$M_{lm}\left(-k^2, t\right) = \frac{-i(2l+1)!!}{c(-ik)^l \sqrt{4\pi(2l+1)\frac{(l+1)}{l}}} \times \int \vec{j}(\vec{r}, t) j_l^*(kr) Y_{llm}^*(\vec{n}) d^3r \tag{2.84}$$

$$\begin{aligned} T_{lm}\left(-k^2, t\right) = {} & \frac{-(2l-1)!!\sqrt{l}}{c(-ik)^{l+1}\sqrt{4\pi(l+1)}} \\ & \times \int \left\{ \sqrt{l}\, j_{l+1}^*(kr)\, Y_{ll+1m}^*(\vec{n}) \sqrt{l+1} \right. \\ & \left. + \left[j_{l-1}^*(kr) - \frac{4\pi(-ikr)^{l-1}}{(2l-1)!!} \right] \times Y_{ll-1m}^*(\vec{n}) \right\} \vec{j}(\vec{r}, t) d^3r \end{aligned} \tag{2.85}$$

where

$$F_{lmk}(\vec{r}) \sim \lim_{r \to 0} \frac{4\pi(ikr)^l}{(2l+1)!!} Y_{lm}(\vec{n}) \tag{2.86}$$

The conjugate of the equations above can be obtained using the following relations:

$$\begin{cases} Q_{lm}^*(-k^2, t) = (-1)^m Q_{l,-m}(-k^2, t) \\ M_{lm}^*(-k^2, t) = (-1)^m M_{l,-m}(-k^2, t) \\ T_{lm}^*(-k^2, t) = (-1)^m T_{l,-m}(-k^2, t) \end{cases} \tag{2.87}$$

In continue, to calculate the radiation intensity arising from the resonators at large distances, one needs to estimate the definitions of the electric and magnetic fields by considering their properties at large distances. To this end, we use the extracted electric and magnetic fields emitted from a scatterer:

$$\begin{aligned} \vec{E}(\vec{r}, t) = {} & \frac{1}{\pi^{3/2}} \\ & \int_0^\infty \left\{ \sum_{l,m} (-i)^{l-1} \left(\frac{\omega}{c}\right)^{l+2} \frac{\sqrt{2l+1}}{(2l+1)!!} \left[\sin(\omega t) \vec{N}_{lm(\omega/c)}^{(-)}(\vec{r}) + \cos(\omega t) \vec{F}_{lm(\omega/c)}^{(-)}(\vec{r}) \right] Q_{lm}\left(-\frac{\omega^2}{c^2}, \omega\right) \right\} d\omega \\ & - \frac{1}{2\pi^2 c} \int_0^\infty \sum_{lm} (-i)^{l-1} \left(\frac{\omega}{c}\right)^{l+1} \frac{\sqrt{4\pi(2l+1)(l+1)}}{\sqrt{l}(2l+1)!!} \left\{ \left[\sin(\omega t) \vec{N}_{lm(\omega/c)}^{(-)}(\vec{r}) + \cos(\omega t) \vec{F}_{lm(\omega/c)}^{(-)}(\vec{r}) \right] \right. \end{aligned}$$

$$\times \frac{\sqrt{l}}{\sqrt{l+1}}\omega Q_{lm}\left(-\frac{\omega^2}{c^2},\omega\right)+\omega\left[\sin(\omega t)\vec{N}^{(0)}_{lm(\omega/c)}(\vec{r})+\cos(\omega t)\vec{F}^{(0)}_{lm(\omega/c)}(\vec{r})\right]M_{lm}\left(-\frac{\omega^2}{c^2},\omega\right)$$

$$+\left[\sin(\omega t)\vec{N}^{(+)}_{lm(\omega/c)}(\vec{r})+\cos(\omega t)\vec{F}^{(+)}_{lm(\omega/c)}(\vec{r})\right]\left[\omega Q_{lm}(0,\omega)+\frac{\omega^2}{c}T_{lm}\left(-\frac{\omega^2}{c^2},\omega\right)\right]\Bigg\}d\omega \tag{2.88}$$

$$\vec{B}(\vec{r},t)=\frac{1}{2\pi^2}$$

$$\int_0^\infty\sum_{l,m}(-i)^{l-1}\left(\frac{\omega}{c}\right)^{l+2}\times\frac{\sqrt{4\pi(2l+1)(l+1)}}{\sqrt{l}(2l+1)!!}\Bigg\{\left[-\sin(\omega t)\vec{F}^{(+)}_{lm(\omega/c)}(\vec{r})+\cos(\omega t)\vec{N}^{(+)}_{lm(\omega/c)}(\vec{r})\right.$$

$$M_{lm}\left(-\frac{\omega^2}{c^2},\omega\right)+\left[-\sin(\omega t)\vec{F}^{(0)}_{lm(\omega/c)}(\vec{r})+\cos(\omega t)\vec{N}^{(0)}_{lm(\omega/c)}(\vec{r})\right]$$

$$\times\left[Q_{lm}(0,\omega)+\frac{\omega^2}{c}T_{lm}\left(-\frac{\omega^2}{c^2},\omega\right)\right]\Bigg\}d\omega \tag{2.89}$$

By calculating the electromagnetic fields in the order of $r\to\infty$, we have:

$$\vec{E}(\vec{r},t)\underset{r\to\infty}{\sim}\frac{1}{r}\sum_{lm}\left(\frac{1}{c}\right)^{l+1}\frac{\sqrt{4\pi(2l+1)(l+1)}}{\sqrt{l}(2l+1)!!}\times\left\{-Q^{(0)(l+1)}_{lm}\left(0,t-\frac{r}{c}\right)\right.$$

$$\times\left[\frac{\sqrt{l}}{\sqrt{2l+1}}\vec{Y}_{ll+1m}(\vec{n})+\frac{\sqrt{l+1}}{\sqrt{2l+1}}\vec{Y}_{ll-1m}(\vec{n})\right]-\sum_{n=0}^{\infty}\frac{1}{n!c^{2n}}M^{(n)(l+2n+1)}_{lm}\left(0,t-\frac{r}{c}\right)\vec{Y}_{llm}(\vec{n})$$

$$+\frac{1}{c}\sum_{n=0}^{\infty}\frac{1}{n!c^{2n}}T^{(n)(l+2n+2)}_{lm}\left(0,t-\frac{r}{c}\right)\times\left[\frac{\sqrt{l}}{\sqrt{2l+1}}\vec{Y}_{ll+1m}(\vec{n})+\frac{\sqrt{l+1}}{\sqrt{2l+1}}\vec{Y}_{ll-1m}(\vec{n})\right] \tag{2.90}$$

$$\vec{B}(\vec{r},t)\underset{r\to\infty}{\sim}\frac{1}{r}\sum_{lm}\left(\frac{1}{c}\right)\frac{\sqrt{4\pi(2l+1)(l+1)}}{\sqrt{l}(2l+1)!!}$$

$$\times\left\{-iQ^{(0)(l+1)}_{lm}\left(0,t-\frac{r}{c}\right)\vec{Y}_{llm}(\vec{n})\times\sum_{n=0}^{\infty}\frac{1}{n!c^{2n}}M^{(n)(l+2n+1)}_{lm}\left(0,t-\frac{r}{c}\right)\vec{Y}_{llm}(\vec{n})\right.$$

$$\left.\times\left[\frac{\sqrt{l}}{\sqrt{2l+1}}\vec{Y}_{ll+1m}(\vec{n})+\frac{\sqrt{l+1}}{\sqrt{2l+1}}\vec{Y}_{ll-1m}(\vec{n})\right]+\frac{1}{c}\sum_{n=0}^{\infty}\frac{1}{n!c^{2n}}T^{(n)(l+2n+2)}_{lm}\left(0,t-\frac{r}{c}\right)\right\} \tag{2.91}$$

Employing (2.90), one can obtain the far-field radiation pattern of an oscillating toroidal dipole as following (Radescu and Vlad 1998; Costescu and Radescu 1987; Arfken and Weber 2001):

$$E(\mathbf{r})=\frac{-i\mu_0c^2k^3}{3\sqrt{2\pi}r}\exp(-ikr)\times\sum_{m=0,\pm1}T_{1m}\left[\mathbf{Y}_{1,2,m}+\sqrt{2}\mathbf{Y}_{1,0,m}\right] \tag{2.92}$$

$$T_{1,\pm1}=\frac{1}{\sqrt{2}}\left(\mp T_x+iT_y\right) \tag{2.93}$$

$$T_{1,0}=T_z \tag{2.94}$$

$$\mathbf{T}=\frac{1}{10c}\int\left[\mathbf{r}(\mathbf{r}\cdot\mathbf{j})-2r^2\mathbf{j}\right]d^3r \tag{2.95}$$

where μ_0 is the magnetic permeability of vacuum, $\mathbf{r}$ is the vector between the location of the dipole moment and the detector, $\mathbf{T}$ is the toroidal dipole mode, and $\mathbf{j}$ is the current density. The total field scattered by an emitter, consists of infinite arrays of resonators, ($\mathbf{E_s}$) can be obtained by the summation of contributions from all dipoles at the position of the observer. We assumed that all excited dipoles are oscillating in-phase and the emitter arrays are adequately smaller than the incident beam wavelength. Then, the complex-valued field emitted by the single emitter at $\mathbf{r}_s$ and detected by the detector at point $\mathbf{r}_d$ is given by (Radescu and Vlad 1998):

$$A^{(s)}_{k,l,m}(\mathbf{r}_d;\ \mathbf{r}_s) = Y_{l,m}\left(\frac{\mathbf{r}_d - \mathbf{r}_s}{|\mathbf{r}_d - \mathbf{r}_s|}\right)\frac{\exp(-ik|\mathbf{r}_d - \mathbf{r}_s|)}{|\mathbf{r}_d - \mathbf{r}_s|} \tag{2.96}$$

in which k is the wave number and Y denotes the spherical harmonics (Arfken and Weber 2001). In this limit, the total field reaching the detector at distance R will be:

$$A^{(d)}_{k,l,m}(x_d, y_d, R) = \sum_n A^{(s)}_{k,l,m}(x_d, y_d, R; \mathbf{r}_n) \tag{2.97}$$

Now, using this assumption that the array lies in the xy-plane at $z = 0$ and the detector is located at $z = R$, we can substitute the sum over the metamolecules of the array with following integral:

$$\mathbf{E_s} = \sum_{\mathbf{r}} \mathbf{E}(\mathbf{r}) \approx \frac{1}{\Delta^2}\int \mathbf{E}(\mathbf{r})d^2r \tag{2.98}$$

where Δ is the area of the metamolecule. After integration, we have:

$$A^{(d)}_{k,l,m}(x_d, y_d, R) \cong \frac{\pi\delta_{m,0}}{ik\Delta^2}\sqrt{\frac{2l+1}{\pi}}\exp(-ikR) \tag{2.99}$$

By considering the propagation direction of radiation and focusing on the far-field component of the radiation for the distances much bigger than the wavelength of the incident wave, one can claim that:

$$I_{l,m} \cong \frac{\pi\delta_{m,0}\left(\hat{\mathbf{R}}.\hat{\mathbf{z}}\right)^l}{ik}\sqrt{\frac{2l+1}{\pi}}\exp(-ikR) \tag{2.100}$$

Then, by substituting (2.92) into (2.98), and utilizing (2.100), we have:

$$\mathbf{E_s} = \frac{\mu_0 c^2 k^2}{4\Delta^2}\sqrt{2}\begin{pmatrix} T_{1,1} - T_{1,-1} \\ i(T_{1,1} - T_{1,-1}) \\ 0 \end{pmatrix}\exp(-ikR) \tag{2.101}$$

Further simplifications on the equation above leads to the final form of the scattered field as (Afanasiev 2001):

$$\mathbf{E_s} = -\frac{\mu_0 c^2 k^2}{2\Delta^2}\mathbf{T}_{\parallel}\exp(-ikR) \tag{2.102}$$

For a particular scatterer, the projected toroidal dipole ($\mathbf{T}_{\parallel}$) is given by $\mathbf{T}_{\parallel} = T_x\hat{\mathbf{x}} + T_y\hat{\mathbf{y}}$, or in general $\mathbf{T}_{\parallel} = \mathbf{T} - (\mathbf{T}\cdot\hat{\mathbf{R}})\,\hat{\mathbf{R}}$. To derive the far-field distribution of other isolated multipoles, one should use the expression for the radiation emitted by multipole sources. Owing to the large number of terms, it would be more efficient to separate the series into different orders of l. Below, we listed the dipolar ($l = 1$), quadrupolar ($l = 2$), and octupolar ($l = 3$) contributions:

$$\mathbf{E}_{(l=1)} \approx \frac{\mu_0 c^2}{3\sqrt{2\pi}}\frac{\exp(-ikR)}{r} \times \sum_{m=0,\pm 1}\Big\{\left(k^2 Q_{1,m} - ik^3 T_{1,m} + ik^5 T^{(1)}_{1,m}\right) \\ \times \left(\mathbf{Y}_{1,2,m} + \sqrt{2}\mathbf{Y}_{1,0,m}\right) + \mathrm{i}\sqrt{3}\left(k^2 M_{1,m} - k^4 T^{(1)}_{1,m}\right) \times \mathbf{Y}_{1,1,m}\Big\} \tag{2.103}$$

$$\mathbf{E}_{(l=2)} \approx \frac{\mu_0 c^2}{10\sqrt{6\pi}}\frac{\exp(-ikR)}{r} \times \sum_{m=0,\pm 1,\pm 2}\Big\{\left(ik^2 Q^{(e)}_{2,m} + k^5 Q^{(T)}_{2,m}\right) \\ \times \left(\sqrt{2}\mathbf{Y}_{2,3,m} + \sqrt{3}\mathbf{Y}_{2,1,m}\right) - \mathrm{i}\sqrt{5}k^3 Q^{(m)}_{2,m}\mathbf{Y}_{2,2,m}\Big\} \tag{2.104}$$

$$\mathbf{E}_{(l=3)} \approx -\frac{\mu_0 c^2 k^4}{15\sqrt{3\pi}}\frac{\exp(-ikR)}{r} \times \sum_{m=0,\pm 1,\pm 2,\pm 3}\Big\{\frac{1}{7}O^{(e)}_{3,m}\left(\sqrt{3}\mathbf{Y}_{3,4,m} + 2\mathbf{Y}_{3,2,m}\right) \\ + \frac{\mathrm{i}}{\sqrt{7}}O^{(m)}_{3,m}\mathbf{Y}_{3,3,m}\Big\} \tag{2.105}$$

and, the total emitted field is given by:

$$\mathbf{E} = \mathbf{E}_{(l=1)} + \mathbf{E}_{(l=2)} + \mathbf{E}_{(l=3)} + \cdots \tag{2.106}$$

It should be underlined that the terms above $l > 3$ (e.g., hexadecapole) can be neglected due to their weak impact on the total emitted field. Ultimately, the far-field for the electric field propagation can be written as (Savinov et al. 2014):

$$\mathbf{E}_s = \frac{\mu_0 c^2}{2\Delta^2}\Big\{-ik\mathbf{p}_{\parallel} + ik\hat{\mathbf{R}} \times \left[\mathbf{m}_{\parallel} - \frac{k^2}{10}\mathbf{m}^{(1)}_{\parallel}\right] - k^2\left[\mathbf{T}_{\parallel} - \frac{k^2}{10}\mathbf{T}^{(1)}_{\parallel}\right] \\ + k^2\left[\mathbf{Q}^{(\mathrm{e})}\cdot\hat{\mathbf{R}}\right]_{\parallel} - \frac{k^2}{2}\hat{\mathbf{R}} \times \left[\mathbf{Q}^{(\mathrm{m})}\cdot\hat{\mathbf{R}}\right]_{\parallel} - \frac{ik^3}{3}\left[\mathbf{Q}^{(\mathrm{T})}\cdot\hat{\mathbf{R}}\right]_{\parallel} \\ - ik^3\left[\left(\mathbf{O}^{(\mathrm{e})}\cdot\hat{\mathbf{R}}\right)\hat{\mathbf{R}}\right]_{\parallel} - \frac{ik^3}{180}\left[\left(\mathbf{O}^{(\mathrm{m})}\cdot\hat{\mathbf{R}}\right)\hat{\mathbf{R}}\right]_{\parallel}\Big\} \times \exp(-ikR) \tag{2.107}$$

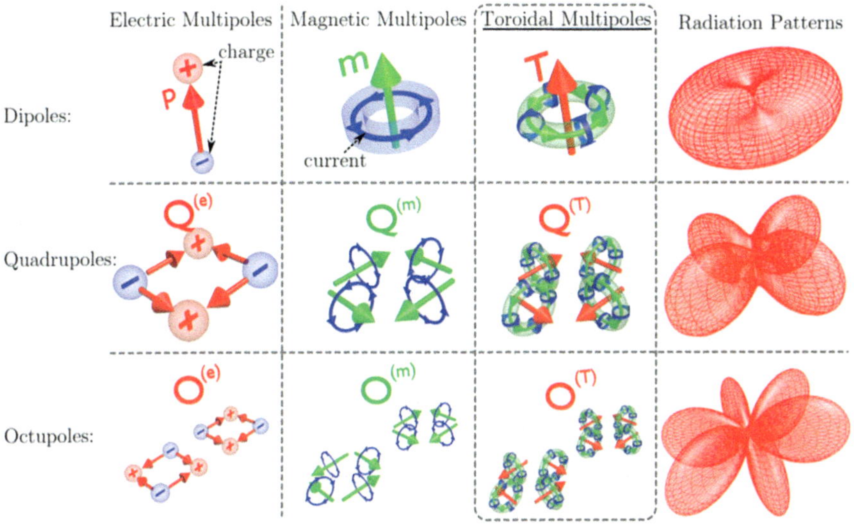

Fig. 2.14 Three families of dynamic multipoles. Artistic representation of charge configurations and far-field radiation patterns of electric, magnetic, and toroidal multipoles. Adapted from (Savinov et al. 2014) with permission. Copyright APS

Here, the terms that are contributed to the emitted field are electric (**p**), magnetic (**m**), and toroidal dipoles (**T**), electric ($\mathbf{Q}^{(e)}$), magnetic ($\mathbf{Q}^{(m)}$), and toroidal ($\mathbf{Q}^{(T)}$) quadrupoles, electric($\mathbf{O}^{(e)}$), magnetic($\mathbf{O}^{(m)}$), and toroidal ($\mathbf{O}^{(T)}$) octupoles. The mean-square radii of toroidal and magnetic dipoles are denoted as $\mathbf{T}^{(1)}$ and $\mathbf{m}^{(1)}$, respectively. It is important to note that to analyze the spectral response of scatterers, the first part of (2.107) should be considered, due to weak far-field radiation contribution of high-order multipoles.

Inducing a dynamic toroidal dipole possesses the creation of a closed-loop configuration of the magnetic fields and currents rotating on the surface of a torus (Fig. 2.14). This graph demonstrates the charge and current configurations for classical electromagnetic and toroidal multipoles, where the far-field radiation pattern of both classical and toroidal multipoles is indistinguishable. Since the electromagnetic manifestations of the toroidal dipole dramatically masks by stronger conventional charge and magnetic multipoles, the observation of toroidal response extremely challenging, and required multipole expansion analysis that will be discussed in the future Chapters.

2.6.3 Dynamic Anapoles

In Sect. 2.6.2, optically driven (dynamic) toroidal dipole is introduced as an independent term in the family of electrodynamic multipole expansion. It is also shown that static toroidal dipoles, known as *static anapoles* described the context of parity

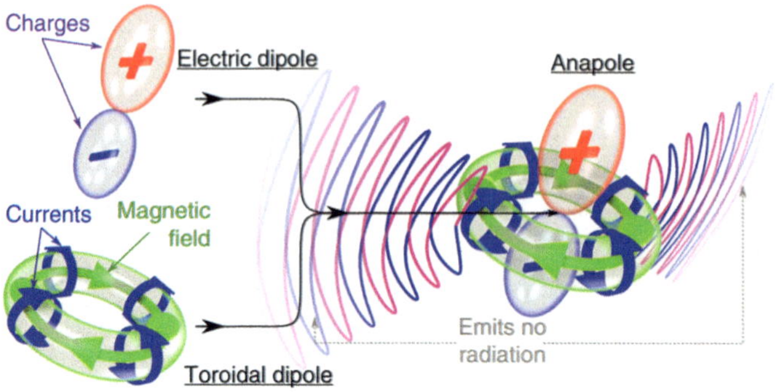

Fig. 2.15 Principle of a dynamic anapole. Anapole is a superposition of electric and toroidal dipoles. Anapole mode arises when the fields radiated by the electric and toroidal dipoles destructively cancel each other. Adapted from (Savinov et al. 2019) with permission. Copyright Springer Nature

violation in nuclear physics, have been observed in magnetism and could be the only possible electromagnetic form factor for dark matter candidate particles (Zel'Dovich 1958; Spaldin et al. 2008; Ho and Scherrer 2013; Gao et al. 2014; Latimer 2017; Alves et al. 2018). In contrast to static anapole, *dynamic anapoles* have been proposed under the toroidal framework, where an electric dipole (a pair of oscillating charges) together with a toroidal dipole (oscillating poloidal current on a torus) form a non-radiating charge-current configuration (Afanasiev and Stepanovsky 1995; Raybould et al. 2017; Baryshnikova et al. 2019; Savinov et al. 2019). In principle, the dynamic anapole state appears at a particular wavelength of oscillations when the fields radiated by the co-located electric and toroidal dipoles destructively cancel each other (Fig. 2.15). The electric and magnetic fields radiated by superposition of these two dipoles are (Basharin et al. 2017; Savinov 2018):

$$E_A = E_p + E_T = \frac{\exp(-ikr + i\omega t)}{r} \left\{ \frac{\boldsymbol{r} \cdot (\boldsymbol{p} - ik\boldsymbol{T}) O(\omega, r)}{c^2 r^2} \boldsymbol{r} - \frac{G(\omega, r)}{c^2} (\boldsymbol{p} - ik\boldsymbol{T}) \right\} \tag{2.108}$$

$$H_A = H_p + H_T = -\frac{\exp(-ikr + i\omega t)}{r} \{\boldsymbol{r} \times (\boldsymbol{p} - ik\boldsymbol{T})\} \frac{ikD(\omega, r)}{cr} \tag{2.109}$$

where $O(\omega, r)$, $G(\omega, r)$, and $D(\omega, r)$ are the vector potentials corresponding to the induced currents (Afanasiev and Stepanovsky 1995).

The far-field emission of an anapole mode is zero, due to the identical radiation pattern of electric and toroidal dipoles, and the detection of a perfect anapole through far-field radiation analysis is not possible, since these nonradiative spectral features do not absorb light. In fact, anapole states can be detected if they weakly couple to the electromagnetic fields from free space. Another alternative technique has been developed based on the slightly off-balance anapole excitations. In this limit, if the anapole states are not perfectly balanced, i.e., the electric dipole emission does not precisely cancel out the toroidal dipole radiation, this results in the creation of a narrow peak in the scattering cross-section. Based on this fact, electromagnetic anapoles were initially detected as narrow transmission peaks in the spectra of a microwave metamaterial (Fedotov et al. 2013). After that, the excitation of dynamic anapoles in artificial media was discussed by developing diverse all-dielectric and metallic architectures at subwavelength scales (Gongora et al. 2017; Yang et al. 2018; Wu et al. 2018; Baryshnikova et al. 2019).

References

G.N. Afanasiev, The electromagnetic field of solenoids with time-dependent currents. J. Phys. A: Math. Gen. **23**, 5755–5764 (1990)

G.N. Afanasiev, Simplest sources of electromagnetic fields as a tool for testing the reciprocity-like theorems. J. Phys. D Appl. Phys. **34**, 539–559 (2001)

G.N. Afanasiev, V.M. Dubovik, Some remarkable charge–current configurations. Phys. Part. Nuclei **29**, 366–391 (1998)

G.N. Afanasiev, M. Nelhiebel, Y.P. Stepanovsky, The interaction of magnetizations with an external electromagnetic field and a time-dependent magnetic Aharonov-Bohm effect. Phys. Script. **54**, 417 (1996)

G.N. Afanasiev, Y.P. Stepanovsky, The electromagnetic field of elementary time-dependent toroidal sources. J. Phys. A: Math. Gen. **28**, 4565–4580 (1995)

A. Alves, A.C.O. Santos, K. Sinha, Collider detection of dark matter electromagnetic anapole moments. Phys. Rev. D **97**, 055023 (2018)

G.B. Arfken, H.J. Weber, *Mathematical Methods for Physicists*, vol. 5 (Harcourt/Academic, Oxford, 2001)

K.V. Baryshnikova, D.A. Smirnova, B.S. Luk'yanchuk, Y.S. Kivshar, Optical anapoles: concepts and applications. Adv. Opt. Mater. **7**, 1801350 (2019)

A.A. Basharin, V. Chuguevsky, N. Volsky, M. Kafesaki, E.N. Economou, Extremely high Q-factor metamaterials due to anapole excitation. Phys. Rev. B **95**, 035104 (2017)

A.D. Boardman, K. Marinov, N. Zheludev, V.A. Fedotov, Nonradiating toroidal structures. SPIE Metamater. **5955**, 595504 (2005)

A. Ceulemans, L.F. Chibotaru, Molecular anapole moments. Phys. Rev. Lett. **80**, 1861–1864 (1998)

A. Costescu, E.E. Radescu, Induced toroid structures and toroid polarizabilities. Phys. Rev. D **35**, 3496 (1987)

V.M. Dubovik, A.A. Cheshkov, Multipole expansion in classic and quantum field theory and radiation. Sov. J. Particles. Nucl. **5**, 318–337 (1974)

V.M. Dubovik, L.A. Tosunyan, V.V. Tugushev, Axial toroidal moments in electrodynamics and solid-state physics. Zh. Eksp. Teor. Fiz **90**, 590–605 (1986)

V.M. Dubovik, V.V. Tugushev, Toroid moments in electrodynamics and solid-state physics. Phys. Rep. **187**(4), 145–202 (1990a)

V.M. Dubovik, V.V. Tugushev, Toroid moments in electrodynamics and solid-state physics. Phys. Rep. **187**, 145–202 (1990b)

V.A. Fedotov, A.V. Rogacheva, V. Savinov, D.P. Tsai, N.I. Zheludev, Resonant transparency and non-trivial non-radiating excitations in toroidal metamaterials. Sci. Rep. **3**, 2967 (2013)

V.V. Flambaum, I.B. Khriplovich, P-odd nuclear forces—a source of parity violation in atoms. Sov. Phys. JETP **52**, 835–839 (1980)

V.V. Flambaum, D.W. Murray, Anapole moment and nucleon weak interactions. Phys. Rev. C **56**, 1641–1644 (1997)

J. Franklin, *Classical Electromagnetism*, 2nd edn (Courier Dover Publications, 2017)
Y. Gao, C.M. Ho, J.S. Robert, Anapole dark matter at the LHC. Phys. Rev. D **89**, 045006 (2014)
B. Gerislioglu, A. Ahmadivand, The observation of high-order charge–current configurations in plasmonic meta-atoms: a numerical approach. Photonics **6**, 43 (2019)
J.S.T. Gongora, A.E. Miroshnichenko, Y.S. Kivshar, A. Fratalocchi, Anapole nanolasers for mode-locking and ultrafast pulse generation. Nat. Commun. **8**, 15535 (2017)
W. Greiner, *Classical Electrodynamics* (Springer Science & Business Media, 2012)
N.A. Gumerov, R. Duraiswami, *Fast multipole methods for the Helmholtz equation in three dimensions* (Elsevier, 2005)
W.C. Haxton, Atomic parity violation and the nuclear anapole moment. Science **275**(5307), 1753–1753 (1997)
S. Hayami, H. Kusunose, Y. Motome, Toroidal order in metals without local inversion symmetry. Phys. Rev. B **90**, 024432 (2014)
C.M. Ho, R.J. Scherrer, Anapole dark matter. Phys. Lett. B **722**, 341–346 (2013)
F. Hong-Jian, L. Fa-Min, *Ab initio* prediction on ferrotoroidic and electronic properties of olivine Li4MnFeCoNiP4O16. Chin. Phys. B **18**, 2481 (2009)
J.D. Jackson, *Classical Electrodynamics*, 3rd edn (Wiley, 1998)
C. Kittel, Theory of the structure of ferromagnetic domains in films and small particles. Phys. Rev. **70**, 965–971 (1946)
D.C. Latimer, Anapole dark matter annihilation into photons. Phys. Rev. D **95**, 095023 (2017)
B. Mettout, P. Tolédano, M. Fiebig, Symmetry replication and toroidic effects in the multiferroic pyroxene $NaFeSi_2O_6$. Phys. Rev. B **81**, 214417 (2010)
I.I. Naumov, L. Bellaiche, H. Fu, Unusual phase transitions in ferroelectric nanodisks and nanorods. Nature **432**, 737–740 (2004)
H.C. Oersted, Experiments on the effect of a current of electricity on the magnetic needles, in *Annals of Philosophy*, vol. 16 (Baldwin, Craddock, Joy, London, 1820), p. 273
B. Ögüt, N. Talebi, R. Vogelgesang, W. Sigle, P.A. Van Aken, Toroidal plasmonic eigenmodes in oligomer nanocavities for the visible. Nano Lett. **12**, 5239–5244 (2012)
N. Papasimakis, V.A. Fedotov, V. Savinov, T.A. Raybould, N.I. Zheludev, Electromagnetic toroidal excitations in matter and free space. Nat. Mater. **15**(3), 263–271 (2016)
S.J. Plimpton, W.E. Lawton, A very accurate test of Coulomb's law of force between charges. Phys. Rev. **50**(11), 1066 (1936)
E.E. Radescu, D.H. Vlad, Angular momentum loss by a radiating toroidal dipole. Phys. Rev. E **57**, 6030 (1998)
T. Raybould, *Toroidal Excitations in Free-Space and Metamaterials* (University of Southampton, Doctoral Thesis, 2017)
T. Raybould, V.A. Fedotov, N. Papasimakis, I. Youngs, N.I. Zheludev, Exciting dynamic anapoles with electromagnetic doughnut pulses. Appl. Phys. Lett. **111**, 081104 (2017)
V. Savinov, Light emission by accelerated electric, toroidal, and anapole dipolar sources. Phys. Rev. A **97**, 063834 (2018)
V. Savinov, V.A. Fedotov, N.I. Zheludev, Toroidal dipolar excitation and macroscopic electromagnetic properties of metamaterials. Phys. Rev. B **89**, 205112 (2014)
V. Savinov, N. Papasimakis, D.P. Tsai, N.I. Zheludev, Optical anapoles. Commun. Phys. **2**, 69 (2019)
J.V. Smith, Magnesium pyroxenes at high temperature: inversion in clinoenstatite. Nature **222**, 256–257 (1969)
N.A. Spaldin, M. Fiebig, M. Mostovoy, The toroidal moment in condensed-matter physics and its relation to the magnetoelectric effect. J. Phys. Condens. Mat. **20**, 434203 (2008)
Y. Tokura, Multiferroics—toward strong coupling between magnetization and polarization in a solid. J. Magn. Magn. Mater. **310**, 1145–1150 (2007)
P. Toledano, D.D. Khalyavin, L.C. Chapon, Spontaneous toroidal moment and field-induced magnetotoroidic effects in $Ba_2CoGe_2O_7$. Phys. Rev. B **84**(9), 094421 (2011)
P. Tolédano, M. Ackermann, L. Bohatý, P. Becker, T. Lorenz, N. Leo, M. Fiebig, Primary ferrotoroidicity in antiferromagnets. Phys. Rev. B **92**(9), 094431 (2015)

B.B. Van Aken, J.-P. Rivera, H. Schmid, M. Fiebig, Observation of ferrotoroidic domains. Nature **449**, 702–705 (2007)

C.S. Wood, S.C. Bennett, D. Cho, B.P. Masterson, J.L. Roberts, C.E. Tanner, C.E. Wieman, Measurement of parity nonconservation and an anapole moment in cesium. Science **275**(5307), 1759–1763 (1997)

P.C. Wu, C.Y. Liao, V. Savinov, T.L. Chung, W.T. Chen, Y.-W. Huang, P.R. Wu, Y.H. Chen, A.Q. Liu, N.I. Zheludev, D.P. Tsai, Optical anapole metamaterial. ACS Nano **12**, 1920–1927 (2018)

Y. Yamaguchi, T. Kimura, Magnetoelectric control of frozen state in a toroidal glass. Nat. Commun. **4**, 2063 (2013)

Y. Yang, V.A. Zenin, S.I. Bozhevolnyi, Anapole-assisted strong field enhancement in individual all-dielectric nanostructures. ACS Photon. **5**, 1960–1966 (2018)

Y.B. Zel'Dovich, Electromagnetic interaction with parity violation. Sov. Phys. JETP **6**(6), 1184–1186 (1958)

I.S. Zheludev, T.M. Perekalina, E.M. Smirnovskaya, S.S. Fonton, Y.N. Yarmukhamedov, Magnetic properties of nickel-boracite iodide. JETP Lett. **20**, 129–130 (1974)

Chapter 3
Expansion of Electromagnetic Multipoles

Abstract This Chapter presents a complete electromagnetic multipole expansion, effective for all point sources in space, including the presence of toroidal moments. To that end, in light of the provided information in Chap. 2, we utilized the solution of inhomogeneous Helmholtz equations to evaluate the electromagnetic field due to alternating poloidal currents in a toroidal solenoid. This solution was obtained through the use of Green's functions and Debye potentials for point sources and fields. The achieved results enabled us to show the physical meaning of unconventional toroidal moments, in comparison to the classical electric and magnetic moments. Besides, the analysis in the long wavelength limit clearly demonstrates that the toroidal moments were neglected previously in the multipole expansion.

3.1 Debye Potentials

By recalling the Maxwell's equations from previous section and rephrasing them as a set of linear equations for harmonic time variations ($e^{-i\omega t}$) with frequency ω, we have:

$$\begin{aligned}
\nabla \cdot \vec{E} &= 4\pi\rho \\
\nabla \times \vec{E} &= \frac{i\omega}{c}\vec{B} \\
\nabla \cdot \vec{B} &= 0 \\
\nabla \times \vec{B} &= \frac{4\pi}{c}\vec{J} - \frac{i\omega}{c}\vec{E}
\end{aligned} \tag{3.1}$$

where ω/c is the wavenumber (k). The equations above can be decoupled by taking the curl of second and fourth relations and using the remaining equations, one can obtain the corresponding inhomogeneous Helmholtz equations for the electric and magnetic fields as following:

$$\left(\nabla^2 + k^2\right)\vec{E}(\vec{r}) = 4\pi\nabla\rho - \frac{4\pi i\omega}{c^2}\vec{J}(\vec{r}) \tag{3.2}$$

A. Ahmadivand et al., *Toroidal Metamaterials*, Engineering Materials,
https://doi.org/10.1007/978-3-030-58288-3_3

$$(\nabla^2 + k^2)\vec{B}(\vec{r}) = -\frac{4\pi}{c}\nabla \times \vec{J}(\vec{r}) \tag{3.3}$$

Here, (3.2) exhibits that the electric field intensity can be defined by the gradient of the charge density and both longitudinal and transverse components of the current density. Conversely, (3.3) shows that the magnetic induction field can be determined by the transverse component of the current density. The solutions of (3.2) and (3.3) can be configured through the use of Green's function for the Helmholtz equation:

$$(\nabla^2 + k^2)\wp_+\left(\vec{r};\vec{r}^{'}\right) = -4\pi\delta\left(\vec{r}-\vec{r}^{'}\right) \tag{3.4}$$

Then, the outgoing wave Green's function and its multipole expansion can be defined by using (Gray 1978):

$$\wp_+\left(\vec{r};\vec{r}^{'}\right) = \frac{e^{ik\left|\vec{r}-\vec{r}^{'}\right|}}{\left|\vec{r}-\vec{r}^{'}\right|} = 4\pi ik\sum_{l=0}^{\infty} j_l(kr_<)h_l^{(1)}(kr_>)$$
$$\sum_{l=0}^{\infty}(-)^m N_{lm}^2 P_l^m\left(\cos\theta^{'}\right)P_l^m(\cos\theta)(2-\delta_{m0})\cos m\left(\varphi-\varphi'\right) \tag{3.5}$$

The solutions of (3.2) and (3.3) involve the integrations of the respective sources and the Green's function:

$$\vec{E}(\vec{r}) = -\int dv'\nabla'\rho\left(\vec{r}^{'}\right)\wp_+\left(\vec{r};\vec{r}^{'}\right) + \frac{i\omega}{c^2}\int dv'\vec{J}\left(\vec{r}^{'}\right)\wp_+\left(\vec{r};\vec{r}^{'}\right) \tag{3.6}$$

$$\vec{B}(\vec{r}) = \int dv'\nabla' \times \vec{J}\left(\vec{r}^{'}\right)\wp_+\left(\vec{r};\vec{r}^{'}\right) \tag{3.7}$$

Applying multipole expansion of the Green's function $\left(\wp_+\left(\vec{r};\vec{r}^{'}\right)\right)$ in (3.5)–(3.7) gives rise to the general and exact expansion of the electromagnetic field. This will also be applied to the toroidal solenoid analysis in the following subsections.

The Debye potentials (Gray 1978; Góngora and Ley-Koo 2006) are useful components for exhibiting the decomposition of the source and force fields into their longitudinal and transverse (toroidal and poloidal) components, including the connection between them. Importantly, gradient of the charge density is a longitudinal field, since its curl is zero. Therefore, the current density is:

$$\begin{aligned}\vec{J}\left(\vec{r}^{'}\right) &= \nabla i^L\left(\vec{r}^{'}\right) + \nabla\times\left(\vec{r}i^T\left(\vec{r}^{'}\right)\right) + \nabla\times\left[\nabla\times\left(\vec{r}\times i^P\left(\vec{r}^{'}\right)\right)\right]\\ &=\nabla i^T\left(\vec{r}^{'}\right) - i\vec{l}i^T\left(\vec{r}^{'}\right) - i\nabla\times i\vec{l}i^P\left(\vec{r}^{'}\right)\end{aligned} \tag{3.8}$$

In the equation above, the corresponding poloidal and toroidal Debye potentials are represented as $-i\nabla^2 i^P(\vec{r})$ and $i^T(\vec{r})$, respectively, and the final response of the equation above motivates the angular momentum operator $\vec{l} = -i\vec{r} \times \nabla$, and the source density can be utilized to fulfill the continuity equation:

$$\nabla.\vec{J}(\vec{r}) = i\omega\rho(\vec{r}) \tag{3.9}$$

It should be noted that this equation contains only the longitudinal component of the current density. In terms of the corresponding Debye potential, (3.9) can be written as:

$$\nabla^2 i^L = i\omega\rho \tag{3.10}$$

This simplified equation demonstrates that the potential $\left(i^L(\vec{r})\right)$ and charge density are related through Poisson's equation. On the other hand, curl of the current density can be attained using triple product and orthogonality of the divergence and angular momentum operators, along with the commutability of Laplace and angular momentum operators:

$$\begin{aligned}\nabla \times \vec{J}(\vec{r}) &= -i\nabla \times \vec{l} i^T(\vec{r}) - i\nabla \times \left[\left(\nabla \times \vec{l}\right) i^P(\vec{r})\right] \\ &= -i\nabla \times \vec{l} i^T(\vec{r}) - i\vec{l}\left[-\nabla^2 i^P(\vec{r})\right]\end{aligned} \tag{3.11}$$

Considering both (3.8) and (3.11) with the symmetric nature of Green's function and hermiticity of gradient and angular momentum operators lead us to rewrite (3.6) and (3.7), and unveil the longitudinal and transverse components of the associated force fields as:

$$\begin{aligned}\vec{E}(\vec{r}) = &-\nabla \int dv'\left[\rho\left(\vec{r}'\right) - \frac{i\omega}{c^2} i^L\left(\vec{r}'\right)\right]\wp_+\left(\vec{r};\vec{r}'\right) \\ &- i\vec{l}\int dv'\left[-\frac{i\omega}{c^2} i^T\left(\vec{r}'\right)\right]\wp_+\left(\vec{r};\vec{r}'\right) \\ &- i\nabla \times \vec{l}\int dv'\left[-\frac{i\omega}{c^2} i^P\left(\vec{r}'\right)\right]\wp_+\left(\vec{r};\vec{r}'\right)\end{aligned} \tag{3.12}$$

$$\begin{aligned}\vec{B}(\vec{r}) = &-i\vec{l} \int dv'\left[-\nabla'^2 i^P\left(\vec{r}'\right)\right]\wp_+\left(\vec{r};\vec{r}'\right) \\ &- i\nabla \times \vec{l} \int dv'\left[i^T\left(\vec{r}'\right)\right]\wp_+\left(\vec{r};\vec{r}'\right)\end{aligned} \tag{3.13}$$

Furthermore, the corresponding Debye potentials can be defined from these equations as:

$$e^L(\vec{r}) = \int dv'\left[\rho\left(\vec{r}'\right) - \frac{i\omega}{c^2} i^L\left(\vec{r}'\right)\right]\wp_+\left(\vec{r};\vec{r}'\right) \tag{3.14}$$

$$e^T(\vec{r}) = \int dv'\left[-\frac{i\omega}{c^2} i^T\left(\vec{r}'\right)\right] \wp_+\left(\vec{r};\vec{r}'\right) \tag{3.15}$$

$$e^P(\vec{r}) = \int dv'\left[-\frac{i\omega}{c^2} i^P\left(\vec{r}'\right)\right] \wp_+\left(\vec{r};\vec{r}'\right) \tag{3.16}$$

$$b^L(\vec{r}) = 0 \tag{3.17}$$

$$b^T\left(\vec{r}'\right) = \int dv'\left[-\nabla'^2 i^P\left(\vec{r}'\right)\right] \wp_+\left(\vec{r};\vec{r}'\right) \tag{3.18}$$

$$b^P\left(\vec{r}'\right) = \int dv'\left[i^T\left(\vec{r}'\right)\right] \wp_+\left(\vec{r};\vec{r}'\right) \tag{3.19}$$

Here, (3.15) and (3.19) indicate that toroidal currents generate toroidal electric and poloidal magnetic fields, while (3.16) and (3.18) exhibit that poloidal currents produce poloidal electric and toroidal magnetic fields. In addition, (3.14) shows that the charge density and longitudinal currents are the sources of the longitudinal electric field.

3.2 Electromagnetic Radiations of Toroidal Solenoids

The unique properties of toroidal excitations were described in the previous Chapter. In this section, we utilize the well-known example of toroidal solenoids to define toroids, with a circular ring cross-section in each meridian plane, and alternating poloidal currents. The curl of the current density is evaluated in (3.7), with the multipole expansion of the Green's function (3.5), to attain the multipole decomposition of the magnetic field. In this framework, the electric field intensity can be assessed by integrating (3.6), where the toroidal nature of the magnetic field and the poloidal feature of the electric field are explicitly exhibited, resulting in the multipole expansions of the related Debye potentials.

3.2.1 The Multipole Decomposition

The toroidal topology with a circular ring cross-section can be accurately defined by its:

1. Inner spherical ring ($r = a$, $\theta_2 < \theta < \theta_1$, φ),
2. Upper conical ring ($a < r < b$, $\vartheta = \theta_1$, φ),
3. Outer spherical ring ($r = b$, $\vartheta_1 < \theta_1 < \vartheta_2$, φ),
4. And lower conical ring ($b > r > a$, $\theta_1 = \vartheta_2$, φ).

The alternating poloidal current ($Ie^{-i\omega t}$) in the toroidal solenoid with N turns has the density of:

$$\vec{J}(\vec{r},t) = \frac{NIe^{-i\omega t}}{2\pi r \sin\theta}\left\{\frac{\hat{r}}{r}[\delta(\theta-\theta_1)-\delta(\theta-\theta_2)][\Theta(r-a)-\Theta(r-b)] + \hat{\theta}[\delta(r-b)-\delta(r-a)][\Theta(\theta-\theta_1)-\Theta(\theta-\theta_2)]\right\} \tag{3.20}$$

Here, the Dirac delta functions indicate the coil elements, where the current fluxes and Heaviside step functions resemble the extent of these elements. Consequently, the curl of the current density is toroidal (in the azimuthal direction) and invariant under rotations around the axis of the moment:

$$\nabla\times\vec{J}(\vec{r}) = \frac{NI\hat{\varphi}}{2\pi r}\left\{\frac{d}{dr}[\delta(r-b)-\delta(r-a)]\frac{\Theta(\theta-\theta_1)-\Theta(\theta-\theta_2)}{\sin\theta} - \frac{1}{r^2}\frac{d}{d\theta}\left[\frac{\delta(\theta-\theta_1)-\delta(\theta-\theta_2)}{\sin\theta}\right][\Theta(\theta-a)-\Theta(\theta-b)]\right\} \tag{3.21}$$

Judicious combination of (3.5), (3.7), and (3.21) leads to the complete multipole expansion of the magnetic field:

$$\begin{aligned}\vec{B}(\vec{r}) = \frac{NI}{2\pi r}\int_0^\infty\int_0^\pi\int_0^{2\pi} r'^2 dr' \sin\theta' d\vartheta' d\varphi' \\ \frac{\hat{\varphi}'}{r'}\left\{\frac{d}{dr'}\left[\delta\left(r'-b\right)-\delta\left(r'-a\right)\right]\frac{\Theta\left(\theta'-\theta_1\right)-\Theta\left(\theta'-\theta_2\right)}{\sin\theta'} - \frac{1}{r'^2}\frac{d}{d\theta'}\left[\frac{\delta\left(\theta'-\theta_1\right)-\delta\left(\theta'-\theta_2\right)}{\sin\theta'}\right]\left[\Theta\left(\theta'-a\right)-\Theta\left(\theta'-b\right)\right]\right\} \\ \{4\pi ik\sum_{l=0}^{\infty} j_l(kr_<)h_l^{(1)}(kr_>) \times \sum_{l=0}^{\infty}(-)^m N_{lm}^2 P_l^m\left(\cos\theta'\right) \\ P_l^m(\cos\theta)(2-\delta_{m0})\cos m\left(\varphi'-\varphi\right)\}\end{aligned} \tag{3.22}$$

where the azimuthal angle integration can be performed through the following relation:

$$\hat{\varphi}' = \hat{\varphi}\cos\left(\varphi'-\varphi\right) - \hat{R}\sin\left(\varphi'-\varphi\right) \tag{3.23}$$

as well as the orthogonality of the cosine and sine functions:

$$\int_0^{2\pi} \hat{\varphi}' \cos m\left(\varphi'-\varphi\right) d\varphi' = \hat{\varphi}\pi\delta_{m1} \tag{3.24}$$

This implies that the magnetic field is also azimuthal and invariant under axial rotations, similar to the toroidal field.

According to the selection rule of (3.24), the polar angle integrations are limited to the terms with $m = 1$ in the sum of (3.22). Thus, the first integral can be expressed as:

$$\int_{\theta_1}^{\theta_2} d\theta' P_l^1\left(\cos\theta'\right) = P_l(\cos\theta_1) - P_l(\cos\theta_2) \tag{3.25}$$

Following from the relation between the associated and ordinary Legendre polynomials, we have:

$$P_l^1(\cos\vartheta) = \sin\theta \frac{d}{d(\cos\theta)} P_l(\cos\theta) \tag{3.26}$$

and the second integral of (3.22), containing the Dirac delta functions, can be solved as below:

$$\begin{aligned} &\int_{\theta_1}^{\theta_2} d\theta' \frac{d}{d\theta'}\left[\sin\theta' P_l^1\left(\cos\theta'\right)\right] \sin\theta' P_l^1\left(\cos\vartheta'\right) \\ &= -\frac{1}{\sin\theta'} \frac{d}{d\theta'}\left[\sin\theta' P_l^1\left(\cos\theta'\right)\right] |\theta' = \theta_1 \\ &\quad + \frac{1}{\sin\theta'} \frac{d}{d\theta'}\left[\sin\theta' P_l^1\left(\cos\theta'\right)\right] |\theta' = \theta_2 \\ &= -l(l+1)[P_l(\cos\theta_1) - P_l(\cos\theta_2)] \end{aligned} \tag{3.27}$$

Combining (3.26) with the differential equation for the ordinary Legendre polynomials leads to:

$$\frac{1}{\sin\theta} \frac{d}{d\theta} \sin\vartheta \frac{d}{d\theta} P_l(\cos\theta) = -l(l+1) P_l(\cos\theta) \tag{3.28}$$

Although in (3.22), the integrations over the radial coordinate are direct, the location difference between the field points must be considered. In light of this, the first integral over the Dirac delta functions can be partially solved as:

$$\int_0^\infty r' dr' \frac{d}{dr'}\left[\delta\left(r' - b\right) - \delta\left(r' - a\right)\right] j_l(kr_<) h_l^{(1)}(kr_>)$$

$$
= \begin{cases} \left(-\frac{d}{dr'}\left[r'h_l^{(1)}(kr')\right]_a^b\right) j_l(kr), & r > a \\ -\frac{d}{dr'}\left[r'h_l^{(1)}(kr')\right]_{r'=b} j_l(kr) + \frac{d}{dr'}\left[r'j_l(kr')\right]_{r'=a} h_l^{(1)}(kr), & a < r < b \\ \left(-\frac{d}{dr'}\left[r'j_l(kr')\right]_a^b\right) h_l^{(1)}(kr), & r > b \end{cases} \tag{3.29}
$$

In the equations above, the second term can be rewritten as:

$$
\int_a^b \frac{d}{dr'} j_l(kr_<) h_l^{(1)}(kr_>)
$$

$$
= \begin{cases} \int_a^b \frac{dr'}{r'} h_l^{(1)}(kr') j_l(kr), & r < a \\ \int_a^r \frac{dr'}{r'} j_l(kr') h_l^{(1)}(kr) + \int_r^b \frac{dr}{r'} h_l^{(1)}(kr') j_l(kr), & a < r < b \\ \int_a^b \frac{dr'}{r'} j_l(kr') h_l^{(1)}(kr), & r > b \end{cases} \tag{3.30}
$$

The substitution of the integrals of (3.24)–(3.29) in (3.22) results in:

$$
\vec{B}(\vec{r}) = \frac{4\pi i k N I}{c} \hat{\varphi} \sum_{l=0}^{\infty} (-) N_{l1}^2 P_l^1(\cos\theta)[P_l(\cos\theta_1) - P_l(\cos\theta_2)]
$$

$$
\times \begin{cases} \left\{-\frac{d}{dr'}\left[r'h_l^{(1)}(kr')\right]_a^b + l(l+1)\int_a^b \frac{d}{dr'} h_l^{(1)}(kr')\right\} jl(kr), & r < a \\ \Big\{-\frac{d}{dr'}\left[r'h_l^{(1)}(kr')\right]_{r'=b} j_l(kr) + \frac{d}{dr'}\left[r'j_l(kr')\right]_{r'=b} h_l^{(1)}(kr') & \\ +l(l+1)\left[\int_a^r \frac{dr'}{r'} j_l(kr') h_l^{(1)}(kr) + \int_r^b \frac{dr'}{r'} h_l^{(1)}(kr') j_l(kr)\right]\Big\}, & a < r < b \\ \left\{-\frac{d}{dr'}\left[r'j_l(kr')\right] h_l^{(1)}(kr')\Big|_a^b + l(l+1)\int_a^b \frac{d}{dr'} j_l(kr')\right\} h_l^{(1)}(kr), & r > b \end{cases} \tag{3.31}
$$

It should be noted that the standard multipole expansion of the electromagnetic field is limited to the region outside the sources, corresponding to $r > b$ in (3.29)–(3.31). In the following subsections, we demonstrated the field in the inner region $r < a$, where there are no point sources, and in the intermediate region $a < r < b$, where the sources are located. This can be accomplished within one and the same calculation by simply differentiating between the different locations of the field point. In addition, the solutions in the source free regions ($r < a$ and $r > b$) are the superposition of the solutions of homogeneous Helmholtz equations, while the solutions in the intermediate region are the nonlinear combinations of the spherical Bessel functions and their derivatives or integrals.

Considering the Debye potentials and using the azimuthal unit vector and associated Legendre polynomials, one can write:

$$\hat{\varphi} P_l^1(\cos\theta) = 2\left(\frac{\pi}{2l+1}\right)^{\frac{1}{2}} \left(-i\vec{l}\right) Y_{l0}(\theta,\varphi) \tag{3.32}$$

Thus, (3.31) can be stated in a toroidal framework as below:

$$\vec{B}(r,\theta,\varphi) = (-il)\sum_{l=1}^{\infty} \frac{8\pi ikNI}{c}(-)\sqrt{\frac{\pi}{2l+1}} N_{l1}^{2}[P_l(\cos\theta_1) - P_l(\cos\theta_2)]Y_{l0}(\theta,\varphi)$$
$$\times \begin{cases} k^2 \int_a^b dr' r' h_l^{(1)}(kr') j_l(kr), & r < a \\ \left\{\left[\frac{d}{dr}(rj_l(kr)) + k^2 \int_a^r dr' r' j_l(kr')\right] h_l^{(1)}(kr) \right. \\ \left. +\left[-\frac{d}{dr}\left(rh_l^{(1)}(kr)\right) + k^2 \int_r^b dr' r' h_l^{(1)}(kr')\right] j_l(kr)\right\}, & a < r < b \\ k^2 \int_a^b dr' r' j_l(kr') h_l^{(1)}(kr) & , r > b \end{cases} \tag{3.33}$$

Moreover, (3.33) can be considered as the multipole expansion of the Debye potential b^T, which can be easily perceived through a simple comparison between (3.13) and (3.18). It should be underlined that owing to hermiticity, the Laplacian in (3.18) can also be performed on the Green's function. According to Helmholtz Eq. (3.4), this gives $-k^2$ times the Green's function plus the point source density term. This can be observed in the radial factors of (3.33). In the set of equations above, the electric field intensity, the current density and the magnetic induction can be defined using (3.1) (fourth term), (3.21), and (3.33), respectively. Here, the electric field intensity is poloidal because of the use of the curl in the toroidal magnetic induction and the poloidal nature of the current density. Besides, the current density can be described in its multipole expansion picture through the use of corresponding representations of the Dirac delta and Heaviside step functions in polar angles, in (3.20).

3.2.2 *The Dynamic Toroidal Multipoles*

In continue, we consider the electric and magnetic fields where $r > b$, in the form that enables the characterization of the multipole expansions of the electromagnetic field of toroidal solenoids. Hence, the definitions for the electric and magnetic fields are:

$$\vec{E}(r > b, \theta, \varphi) = \sum_{l=1}^{\infty} \psi_{l0} \frac{i}{k} \nabla \times \vec{l} h_l^{(1)}(kr) Y_{l0}(\theta, \varphi) \tag{3.34}$$

$$\vec{B}(r > b, \theta, \varphi) = \sum_{l=1}^{\infty} \psi_{l0} \vec{l} h_l^{(1)}(kr) Y_{l0}(\theta, \varphi) \tag{3.35}$$

where

$$\psi_{l0} = -\frac{2\pi k^3 NI}{c} \sqrt{\frac{2l+1}{\pi}} \frac{1}{l(l+1)} [P_l(\cos\theta_1) - P_l(\cos\theta_2)] \int_a^b dr' r' j_l(kr') \tag{3.36}$$

For the purpose of evaluation of electric and magnetic fields, we write down the general form of multipole expansion of electromagnetic field as following (Gray 1978):

$$\vec{E}(\vec{r}) = \sum_l \sum_m \left\{ \psi_{lm}^E \left(\frac{i}{k} \right) \nabla \times \vec{l} h_l^{(1)}(kr) Y_{lm}(\theta, \varphi) + \psi_{lm}^M \vec{l} h_l^{(1)}(kr) Y_{lm}(\theta, \varphi) \right\} \tag{3.37}$$

$$\vec{B}(\vec{r}) = \sum_l \sum_m \left\{ \psi_{lm}^E \vec{l} h_l^{(1)}(kr) Y_{lm}(\theta, \varphi) + \psi_{lm}^M \frac{(-i)}{k} \nabla \times \vec{l} h_l^{(1)}(kr) Y_{lm}(\theta, \varphi) \right\} \tag{3.38}$$

Here, the dynamic multipole moments can be described as (Jackson 2007):

$$\psi_{lm}^E = -\frac{4\pi i k^2}{cl(l+1)} \left[\int dv' j_l(kr') Y_{lm}^*(\theta', \varphi') \right] \left\{ ik\vec{r}'.\vec{J}(\vec{r}') - c\left(2 + \vec{r}'.\nabla\right)\rho(\vec{r}') \right\} \tag{3.39}$$

$$\psi_{lm}^M = \frac{4\pi i k^2}{cl(l+1)} \int dv' j_l(kr') Y_{lm}^*(\theta', \varphi') \vec{r}'.\nabla' \times \vec{J}(\vec{r}') \tag{3.40}$$

where ψ_{lm}^E and ψ_{lm}^M represent the dynamic electric and magnetic multipole moments, respectively. In theory, (3.34)–(3.36) and (3.37)–(3.40) can be compared straightforwardly. Since ψ_{lm}^M disappears at the source level due to poloidal character of the alternating current (from (3.20)), the poloidal component of the magnetic field is missing in (3.35) and the toroidal component of the electric intensity field is missing in (3.37). In addition, the disappearance of the integrand in (3.40) follows from orthogonality of the radial vector and curl of the current density (3.21). On the other hand, the dynamic electric multipole moments in (3.39) are defined by the radial components of the current density and of the gradient of the charge density. Eventually, the expansions of (3.34)–(3.36) correspond to the dynamic transverse electric multipole moments associated with the radial component of the poloidal current density in (3.40) with $\rho = 0$.

To finalize the characterization of the multipole expansion of the electromagnetic field of the toroidal solenoid, one can add the superscript E in (3.36)–(3.38). However, this leads to the disappearance of the toroidal moments. Adding the correct superscripts in (3.36)–(3.38) and following a similar approach mentioned in Ref. (Dubovik and Cheshkov 1975), (3.40) gives an exact relationship between the dynamic multipole moments of electric, toroidal, and charge types:

$$\psi^E_{lm}(k) = \psi^T_{lm}(k) + \psi^Q_{lm}(k) \tag{3.41}$$

The differences and similarities between these types of dynamic multipole moments can be understood through separating the source terms of inhomogeneous Helmholtz equations (see (3.2) and (3.3)). In (3.39) and (3.40), it is easy to see that the source components are the radial factors of the respective sources.

We initially consider the unambiguous case of the magnetic moments, in which the corresponding source factor in the integrand of (3.40) can be expressed in terms of the decomposition of the current density provided in (3.8):

$$\vec{r}'.\left[\nabla' \times \vec{J}\left(\vec{r}'\right)\right] = \left[\vec{r}' \times \nabla'\right].\vec{J}\left(\vec{r}'\right) = \vec{l}^2 i^T \tag{3.42}$$

where the dot and cross are exchanged in the triple scalar product, and later, the orthogonality of the operators is applied. (3.42) demonstrates that the dynamic magnetic multipole moments only depend on the toroidal component of the current. Besides, for the electric multipole moments, this comprises both longitudinal and poloidal components of the source:

$$\vec{r}'.\left[-\nabla'\rho\left(\vec{r}'\right) + \frac{ik}{c}\vec{J}\left(\vec{r}'\right)\right] = -\vec{r}'.\nabla'\left[\rho\left(\vec{r}'\right) - \frac{ik}{c}i^L\left(\vec{r}'\right)\right] + i\vec{l}^2 i^P\left(\vec{r}'\right) \tag{3.43}$$

Bearing in mind that the fundamental parts of the multipole expansions are the eigenfunctions of ∇^2 and $\vec{l}^2$ operators, it is obvious that (3.40) is related to (3.19) and (3.41) is correlated with both (3.14) and (3.18).

3.2.3 *The Long Wavelength Regime*

Assuming that the sources are confined in an area with a radial range smaller than the wavelength of incidence, one can approximate the spherical Bessel functions in (3.32)–(3.40) through the dominant terms of their power series expansions near the origin. In addition, the weight functions in (3.39) and (3.40) are associated with the static multipole moments defined by the powers of the radial coordinates as following:

$$\psi_{lm}^{E}(k \to 0) = -\frac{4\pi i k^{l+2}}{cl(l+1)(2l+1)!!} \int dv' r'^{l} Y_{lm}^{*}(\theta', \varphi')$$
$$\left[ik\vec{r}'.\vec{J}\left(\vec{r}'\right) - c(2+\vec{r}.\nabla')\rho\left(\vec{r}'\right)\right]$$
$$= -\frac{4\pi i k^{l+2}}{cl(2l+1)!!}\left[\frac{1}{l+1}\int dv' r'^{l} Y_{lm}^{*}(\theta', \varphi') ik\vec{r}'.\vec{J}\left(\vec{r}'\right)\right.$$
$$\left. + c\int dv' r'^{l} Y_{lm}^{*}(\theta', \varphi')\rho\left(\vec{r}'\right)\right] M_{lm} \tag{3.44}$$

$$\psi_{lm}^{M}(k \to 0) = \frac{4\pi i k^{l+2}}{cl(l+1)(2l+1)!!}$$
$$\int dv' r'^{l} Y_{lm}^{*}(\theta', \varphi')\vec{r}'.\nabla' \times \vec{J}\left(\vec{r}'\right)$$
$$= \frac{4\pi i k^{l+2}}{l(2l+1)!!} M_{lm} \tag{3.45}$$

The part associated with the radial component of the current density in (3.44) can be neglected in the long wavelength approximation. Now, the dynamic multipole moments in (3.39)–(3.41) can be normalized to have the time dependent form factors:

$$M_{lm}\left(-k^2, t\right) = \frac{l(2l+1)!!}{4\pi i k^{l+2}} \psi_{lm}^{M}(k) e^{-i\omega t} \to M_{lm} e^{-i\omega t} \tag{3.46}$$

$$E_{lm}\left(-k^2, t\right) = \frac{cl(2l+1)!!}{4\pi k^{l+1}} \psi_{lm}^{E}(k) e^{-i\omega t} \tag{3.47}$$

$$Q_{lm}\left(-k^2, t\right) = -\frac{l(2l+1)!!}{4\pi i k^{l+2}} \psi_{lm}^{Q}(k) e^{-i\omega t} \to Q_{lm} e^{-i\omega t} \tag{3.48}$$

$$T_{lm}\left(-k^2, t\right) = \frac{l(2l+1)!!}{4\pi k^{l+3}} \psi_{lm}^{T}(k) e^{-i\omega t} \tag{3.49}$$

Based on these facts, Eq. 3.41 can be written as:

$$E_{lm}\left(-k^2, t\right) = k^2 T_{lm}\left(-k^2, t\right) + \dot{Q}_{lm}\left(-k^2, t\right) \tag{3.50}$$

In general, standard systems consisting of electromagnetic sources and fields can be defined in terms of magnetic form factors (see (3.46)), transverse electric form factors (see (3.47)), and charge form factors (see (3.48)). Beyond that, based on the long wavelength approximation, (3.50) can be simplified as.

$$E_{lm}\left(-k^2, t\right) = \dot{Q}_{lm}\left(-k^2, t\right) \tag{3.51}$$

Importantly, for the systems with poloidal currents (e.g., toroidal solenoid), (3.51) is not valid. Instead of the transverse electric components, it is more convenient to consider the toroidal components from (3.49), which are independent from the magnetic and charge elements.

3.2.4 The Magnetostatic Regime

The limit of stationary currents with frequency $\omega \to 0$ in toroidal solenoids is associated with the magnetostatic regime. Thus, the fourth equation in (3.1) becomes Ampère's law, and (3.3) and (3.4) become Poisson's equation. In addition, the Green's function of (3.5) converts to the Coulomb potential, and finally, the spherical Bessel functions are substituted by their power approximations.

By taking up the problem at the level of magnetic field in (3.31), one can write:

$$j_l(kr_<)h_l^{(1)}|_{k\to 0} - \frac{ir_<^l}{k(2l+1)r_<^{l+1}} \tag{3.52}$$

where i is a complex number. For the regions without source ($r < a$ and $r > b$), the radial integrals are finite, but they can be multiplied by the factor of k^2, which disappears in the static limit. In the notation of long wavelength limit, the toroidal components are finite, but the toroidal moments vanish in the inner and outer regions that bounds the toroid. To this end, the magnetic induction becomes ineffective in both regions, and for the region where the sources are located, the radial factor is:

$$\left(\frac{d}{dr}\left(r^{l+1}\right)\right)\frac{1}{r^{l+1}} - \left(\frac{d}{dr}\left(\frac{1}{r^l}\right)\right)r^l = \frac{2l+1}{r} \tag{3.53}$$

The coefficient in the numerator of (3.53) will cancel the factor in the denominator of (3.52). Then, (3.31) can be expressed as:

$$\begin{aligned}\vec{B}(a < r < b, \theta, \varphi) &= \frac{4\pi NI}{cr}\hat{\varphi}\sum_{l=1}^{\infty}(-)\frac{2l+1}{4\pi}P_l^1(\cos\theta)[P_l(\cos\theta_1) - P_l(\cos\theta_2)] \\ &= \frac{2NI}{cr\sin\theta}(\Theta(\theta-\theta_1) - \Theta(\theta-\theta_2))\end{aligned} \tag{3.54}$$

In the equation above, the explicit value of the normalization constant was substituted, and finally, the sum is defined with the difference of the Heaviside step functions in the polar angle, which follows from the completeness of the orthonormal Legendre basis:

$$\sum_{l=0}^{\infty}\frac{2l+1}{2}P_l(\cos\theta)P_l(\cos\theta_i) = \delta(\cos\theta - \cos\theta_i) = -\frac{\delta(\theta-\theta_i)}{\sin\theta} \tag{3.55}$$

and

$$\Theta(\theta-\theta_i)=\int_0^{\theta} d\theta'\delta(\theta-\theta_i)=-\sum_{l=0}^{\infty}\frac{2l+1}{2}\int_0^{\theta} d\theta' \sin\theta' P_l\left(\cos\theta'\right)P_l(\cos\theta_i)$$
$$=-\sum_{l=1}^{\infty}\frac{2l+1}{2l(l+1)}\sin\theta P_l^1(\cos\theta)P_l(\cos\theta_i) \quad (3.56)$$

Here, the magnetic induction is azimuthal and vanishes outside the solenoid, and inversely proportional to the radial distance from the axis in the interior. In addition, noticeably, the toroidal moments between two spheres are not equal to zero, and the summation of all multipole components of the field was carried out in (3.54).

3.2.5 The Point Source Regime

(3.34) and (3.35) for the electromagnetic field of toroidal solenoids correspond to the transverse electric fields of (3.37) and (3.38), respectively. This accompanies with the identification of $\psi^E=\psi^{\mathrm{T}}$, following from (3.41), since the charge moments disappear in this limit. Owing to the identical radiation characteristics of the transverse electric multipole fields, the polarization and angular distribution of the radiation of each multipole components of the field is the same (Gray 1978; Jackson 2007). The difference here is in the amplitudes given by (3.36), which is in contrast with the conventional case dominated by the charge moments in (3.44).

For the solenoid configuration with small dimensions compared to the wavelength of the incidence (i.e., $kb \ll 1$), ψ_{l0} in (3.36) becomes:

$$\psi_{l0}^T(kb<<1)=-\frac{2NI}{c}\frac{\sqrt{\pi(2l+1)}}{l(l+1)}$$
$$[P_l(\cos\theta_1)-P_l(\cos\theta_2)]\frac{k\left[(kb)^{l+2}-(ka)^{l+2}\right]}{l+2} \quad (3.57)$$

Since the ratio of the moments of two consecutive multipoles follows $kb \ll 1$, the element with the lowest multipolarity is the dominant term. In this case, the most dominant moment is the toroidal dipole with $l=1$.

On the other hand, if one compares (3.52) with the corresponding charge multipole moment in (3.44), an extra factor of k for the toroidal moments can be obtained. This can be translated into an additional factor of ω^2 in the radiated power from a toroidal multipole relative to that of the corresponding electric multipole. Consequently, the radiated power from a toroidal solenoid (i.e., toroidal dipole) drives as ω^6, which is in contrast with the well-known ω^4 dependence of classical electric and magnetic dipoles. In fact, the toroidal and electric multipoles of a particular order possess the identical angular momentum and parity properties, hence, the concurrent presence of both types of multipoles yields a frequency dependence of the radiated power that is more complex than the corresponding dependence for each individual type.

To conclude, the complete multipole expansion of the electromagnetic field arising from alternating poloidal currents in toroidal solenoids was constructed. This was performed by assuming that the field exhibits the existence of toroidal multipole moments and termed at all points in space. It is shown that, while poloidal currents in toroidal solenoids hold vanishing magnetic and charge multipole moments, toroidal moments resulting from the poloidal currents, just as magnetic (poloidal) moments resulting from the moments of toroidal currents.

References

V.M. Dubovik, A.A. Cheshkov, Multipole expansion in classical and quantum field theory and radiation. Sov. J. Part. Nucl **5**, 318–337 (1975)

C.G. Gray, Multipole expansions of electromagnetic fields using Debye potentials. Am. J. Phys. **46**, 169–179 (1978)

T. Góngora, E. Ley-Koo, Complete electromagnetic multipole expansion including toroidal moments. Rev. Mex. Fis. E **52**, 177–181 (2006)

J.D. Jackson, *Classical Electrodynamics*. (Wiley, 2007)

Chapter 4
Physical Mechanism Behind the Toroidal Multipoles

Abstract In this chapter, we presented a set of calculations for the radiation intensity, angular momentum loss, and recoil force of the most general type of source, in terms of electric, magnetic, and toroidal multipole moments. In these calculations, we considered a set of studies based on radii of any multipolarity and an arbitrary time dependence. The results are articulated in terms of time derivatives of the multipole moments and mean radii of the associated distributions. To that end, we recalled the equations for the description of electromagnetic multipoles as well as dynamic toroidal moments from Chap. 2. Besides, we employed the classical electrodynamics framework to obtain the rate of angular momentum loss of a time-dependent toroidal dipole, which was derived by Radescu and Vlad (Phys Rev E 57(5):6030, 1998); Radescu and Vaman (Phys Rev E 65:046609, 2002), in connection with a forced precession of the toroidal dipole around a particular axis.

4.1 Defining the Problem

Considering the classical electric and magnetic multipole moments and their distributions, the problem of the absence of toroidal contributions from the radiation intensity, angular momentum, and angular momentum loss, within classical electrodynamics framework, can be solved for any multipole order l. Using the multipole analysis from Chap. 3, including the corresponding notations from Chap. 2, and employing the proposed approach by Dubovik and Tugushev (1990), we provided the calculation of radiation intensity, angular momentum loss, and recoil force of the most general source. This includes all types of electric, magnetic, and toroidal moments, as well as distributions of any multipolarity order and an arbitrary time dependence. The results are described in terms of time derivatives of the mean-square radii of any order n, in which $n = 0$ order implies the corresponding multipole moment itself.

To start with, we recall the charge (2.69) and current (2.70) densities, as well as the set of solutions (2.79)–(2.82) to the basis vector function $\vec{F}^{(\lambda)}_{lmk}(\vec{r})$. Likewise, by considering the electromagnetic and toroidal multipolar form factors (2.83)–2.85), we rephrase them in the limit of $k^2 \to 0$. In this regime, using the electric form factor $Q_{lm}(-k^2, t)$, the electric multipole moments $Q_{lm}(t)$ can be written as:

A. Ahmadivand et al., *Toroidal Metamaterials*, Engineering Materials,
https://doi.org/10.1007/978-3-030-58288-3_4

$$Q_{lm}(t) = Q_{lm}(0, t) = \frac{2\sqrt{\pi}}{\sqrt{2l+1}} \int r^l Y^*_{lm}(\vec{n}) \rho(\vec{r}, t) d^3r \tag{4.1}$$

Rewriting the form factor $Q_{lm}(-k^2, t)$ in Taylor series, one can obtain the mean $2n$-order radii of the $2^{\,l}$-pole charge distribution as:

$$Q_{lm}(-k^2, t) = \sum_{n=0}^{\infty} \frac{(-k^2)^n}{n!} Q^{[n]}_{lm}(t) \tag{4.2}$$

where

$$\begin{aligned} Q^{[n]}_{lm}(t) &= \frac{d^n}{d(-k^2)^n} Q_{lm}(-k^2, t)|_{k^2=0} \\ \bar{r}^{2n}_{lm}(t) &= \frac{2^n(2l+2n+1)!!}{(2l+1)!!} Q^{[n]}_{lm}(t) \\ &= \frac{2\sqrt{\pi}}{\sqrt{2l+1}} \int r^{l+2n} Y^*_{lm}(\vec{n}) \rho(\vec{r}, t) d^3r \end{aligned} \tag{4.3}$$

To extract the proper description related to the high-order derivatives of the form factors, one may use:

$$\frac{d^n}{d(-k^2)^n} \left[\frac{j^*_l(kr)}{k^l} \right] = \frac{(ir)^n}{2^n} \frac{j^*_{l+n}(kr)}{k^{l+n}} \tag{4.4}$$

As mentioned above, radii of zero order ($n = 0$), i.e., the first term in (4.2), are just the multipole moments themselves. This can be mathematically expressed as: $\bar{r}^0_{lm}(t) = Q_{lm}(k^2 = 0, t) = Q^{[0]}_{lm}(t) = Q_{lm}(t)$. Similar approach can also be applied to the magnetic and toroidal multipole form factors:

$$\begin{aligned} M_{lm}(-k^2, t) &= \sum_{n=0}^{\infty} \frac{(-k^2)^n}{n!} M^{[n]}_{lm}(t) \\ M^{[n]}_{lm}(t) &= \frac{d^n}{d(-k^2)^n} M_{lm}(-k^2, t)|_{k^2=0} \end{aligned} \tag{4.5}$$

$$\begin{aligned} T_{lm}(-k^2, t) &= \sum_{n=0}^{\infty} \frac{(-k^2)^n}{n!} T^{[n]}_{lm}(t) \\ T^{[n]}_{lm}(t) &= \frac{d^n}{d(-k^2)^n} T_{lm}(-k^2, t)|_{k^2=0} \end{aligned} \tag{4.6}$$

Now, the multipole magnetic and toroidal moments can be written as:

$$M_{lm}(t) = M^{[0]}_{lm}(t) = M_{lm}(0, t) = \frac{2\sqrt{\pi}}{c(l+1)\sqrt{2l+1}} \int r^l \left[\vec{r} \times \vec{j}(\vec{r}, t) \right] \cdot \vec{\nabla} Y^*_{lm}(\vec{n}) d^3r$$

$$= -\frac{i}{c}\frac{2\sqrt{\pi l}}{c(l+1)\sqrt{2l+1}}\int r^l Y^*_{llm}(\vec{n})\cdot \vec{j}(\vec{r},t)d^3r \tag{4.7}$$

$$T_{lm}(t) = T^{[0]}_{lm}(t) = T_{lm}(0,t) = -\frac{\sqrt{\pi l}}{c(2l+1)}\int r^{l+1}\left\{\vec{Y}^*_{ll-1m}(\vec{n}) + \frac{2\sqrt{l/(l+1)}}{2l+3}\vec{Y}^*_{ll+1m}\right\}\cdot \vec{j}(\vec{r},t)d^3r \tag{4.8}$$

The radii of various $2n$ order are related to the derivatives of the corresponding form factors of the magnetic case by the following equation:

$$\begin{aligned}\bar{\rho}^{2n}_{lm}(t) &= \frac{2^n(2l+2n+1)!!}{(2l+1)!!}M^{[n]}_{lm}(t)\\ &= -\frac{i}{c}\frac{2\sqrt{\pi l}}{\sqrt{(l+1)(2l+1)}}\int r^{2n+l}\vec{Y}^*_{llm}(\vec{n})\cdot\vec{j}(\vec{r},t)d^3r\end{aligned} \tag{4.9}$$

$$\bar{\rho}^{0}_{lm} = M_{lm}(k^2=0,t) = M^{[0]}_{lm}(t) = M_{lm}(t) \tag{4.10}$$

On the other hand, for the toroidal case, the form factor above can be rephrased as:

$$\begin{aligned}\bar{R}^{2n}_{lm}(t) &= \frac{2^n(2l+2n+1)!!}{(2l+1)!!}T^{[n]}_{lm}(t)\\ &= -\frac{1}{c(2l+1)}\frac{2\sqrt{\pi l}}{\sqrt{(l+1)}}\int r^{l+2n+l}\left\{\frac{\sqrt{l}}{(2l+2n+3)}\vec{Y}^*_{ll+1m}(\vec{n})\right.\\ &\quad\left.+\frac{\sqrt{l+1}}{2(n+1)}\vec{Y}^*_{ll-1m}(\vec{n})\right\}\vec{j}(\vec{r},t)d^3r\end{aligned} \tag{4.11}$$

$$\bar{R}^{0}_{lm} = T_{lm}(k^2=0,t) = T^{[0]}_{lm}(t) = T_{lm}(t) \tag{4.12}$$

The following form factors for electrical, magnetic, and toroidal moments are functions with two variables: $Q_{lm}(-k^2,t)$, $M_{lm}(-k^2,t)$, and $T_{lm}(-k^2,t)$. Instead, it would be convenient to describe this information in terms of the electromagnetic framework of the considered system regarding the mean-square radii of various types of electric ($\bar{r}^{2n}_{lm}(t)$), magnetic ($\bar{\rho}^{2n}_{lm}(t)$), and toroidal ($\bar{R}^{2n}_{lm}(t)$) form factors, multipolarity (l), and orders ($2n$). All these radii are still functions of time and their derivatives of diverse orders. Since up to numerical factors, all these radii are themselves derivatives of the form factors $Q_{lm}(-k^2,t)$, $M_{lm}(-k^2,t)$, and $T_{lm}(-k^2,t)$ of various orders (n) with respect to ($-k^2$) at $k^2=0$, we will consider the double-superscript quantities, $Q^{(n)(\upsilon)}_{lm}(0,t)$, $M^{(n)(\upsilon)}_{lm}(0,t)$, and $T^{(n)(\upsilon)}_{lm}(0,t)$:

$$Q_{lm}^{(n)(\upsilon)}(0,t) = \frac{d^\upsilon}{dt^\upsilon}\left[\frac{d^n}{d(-k^2)^n} Q_{lm}(-k^2,t)\big|_{k^2=0}\right] \tag{4.13}$$

$$M_{lm}^{(n)(\upsilon)}(0,t) = \frac{d^\upsilon}{dt^\upsilon}\left[\frac{d^n}{d(-k^2)^n} M_{lm}(-k^2,t)\big|_{k^2=0}\right] \tag{4.14}$$

$$T_{lm}^{(n)(\upsilon)}(0,t) = \frac{d^\upsilon}{dt^\upsilon}\left[\frac{d^n}{d(-k^2)^n} T_{lm}(-k^2,t)\big|_{k^2=0}\right] \tag{4.15}$$

in which the first superscript (n) indicates the order of derivation with respect to $(-k^2)$ at $k^2 = 0$, while the second (υ) represents the order of derivation with respect to time (t) of the corresponding form factor. The correlations between these double-superscript quantities and the derivatives (on the order of υ) with respect to time of the (order of n) mean-square radii of electric, magnetic, and toroidal types are:

$$Q_{lm}^{(n)(\upsilon)}(0,t) = \frac{(2l+1)!!}{2^n(2l+2n+1)!!}\frac{d^\upsilon}{dt^\upsilon}\bar{r}_{lm}^{2n}(t) \tag{4.16}$$

$$M_{lm}^{(n)(\upsilon)}(0,t) = \frac{(2l+1)!!}{2^n(2l+2n+1)!!}\frac{d^\upsilon}{dt^\upsilon}\bar{\rho}_{lm}^{2n}(t) \tag{4.17}$$

$$T_{lm}^{(n)(\upsilon)}(0,t) = \frac{(2l+1)!!}{2^n(2l+2n+1)!!}\frac{d^\upsilon}{dt^\upsilon}\bar{R}_{lm}^{2n}(t) \tag{4.18}$$

The quantities above (4.16)–(4.18) yield the necessary information about the multipole content of the source and it will be in terms of these form factors. Next, these components will be employed for expressing the radiation intensity, angular momentum loss, and recoil force.

4.1.1 Potentials and Fields of a General Source

In this section, we briefly represent and explain the potentials and fields of a particular source to utilize in analyzing the radiation intensity, angular momentum loss, and recoil force. To obtain the fields created by a general distribution of charges and currents, one should recall the continuity relation from (2.68), which consists of current and charge densities. Using this assumption, the retarded scalar $\varphi(\vec{r},t)$ and vector $\vec{A}(\vec{r},t)$ potentials can be written as:

$$\varphi(\vec{r},t) = \int \frac{\rho\left(\vec{r}',t-\frac{|\vec{r}-\vec{r}'|}{c}\right)}{|\vec{r}-\vec{r}'|}d^3\vec{r}' \tag{4.19}$$

$$\vec{A}(\vec{r},t) = \int \frac{\vec{j}\left(\vec{r}',t-\frac{|\vec{r}-\vec{r}'|}{c}\right)}{|\vec{r}-\vec{r}'|}d^3\vec{r}' \tag{4.20}$$

to calculate the fields through the following famous equations:

$$\vec{E}(\vec{r},t) = -\frac{1}{c}\frac{\partial \vec{A}}{\partial t} - \vec{\nabla}\varphi(\vec{r},t) \tag{4.21}$$

$$\vec{B}(\vec{r},t) = \vec{\nabla} \times \vec{A}(\vec{r},t) \tag{4.22}$$

By having the charge (2.69) and current (2.70) densities in time domain, to obtain the electric, magnetic, and toroidal form factors of sources, one needs to define the Fourier transforms of charge density:

$$\rho(\vec{r},\omega) = \int_0^\infty \sin(\omega t)\rho(\vec{r},t)dt$$

$$\rho(\vec{r},t) = \frac{2}{\pi}\int_0^\infty \sin(\omega t)\rho(\vec{r},\omega)d\omega$$

$$\rho(\vec{r},\omega) = \frac{1}{4\pi^3}\sum_{l,m,k}(-ik)^l\frac{\sqrt{\pi(2l+1)}}{(2l+1)!!}\left[Q_{lm}(-k^2,\omega)F_{lmk}(\vec{r})\right] \tag{4.23}$$

with

$$Q_{lm}(-k^2,\omega) = \int_0^\infty \sin(\omega t)Q_{lm}(-k^2,t)dt$$

$$Q_{lm}(-k^2,t) = \frac{2}{\pi}\int_0^\infty \sin(\omega t)Q_{lm}(-k^2,\omega)d\omega \tag{4.24}$$

and current density:

$$\vec{j}(\vec{r},\omega) = \int_0^\infty \cos(\omega t)\vec{j}(\vec{r},t)dt$$

$$\vec{j}(\vec{r},t) = \frac{2}{\pi}\int_0^\infty \cos(\omega t)\vec{j}(\vec{r},\omega)d\omega$$

$$\vec{j}(\vec{r},\omega) = \frac{1}{4\pi^3}\sum_{l,m,k}(-ik)^{l-1}\frac{\sqrt{\pi(2l+1)(l+1)}}{\sqrt{l}(2l+1)!!}\Big\{\Big[ckM_{lm}(-k^2,\omega)F^{(0)}_{lmk}(\vec{r})\Big]$$

$$+\left[\dot{Q}_{lm}(0,\omega) + ck^2T_{lm}(-k^2,\omega)\right]F^{(+)}_{lmk}(\vec{r})$$

$$+\sqrt{\frac{l}{l+1}}\dot{Q}_{lm}(-k^2,\omega)\,F^{(-)}_{lmk}(\vec{r})\Big\} \tag{4.25}$$

with

$$\begin{aligned}
M_{lm}(-k^2,\omega) &= \int_0^\infty \cos(\omega t) M_{lm}(-k^2,t)dt \\
M_{lm}(-k^2,t) &= \frac{2}{\pi}\int_0^\infty \cos(\omega t) M_{lm}(-k^2,\omega)d\omega
\end{aligned} \tag{4.26}$$

$$\begin{aligned}
T_{lm}(-k^2,\omega) &= \int_0^\infty \cos(\omega t) T_{lm}(-k^2,t)dt \\
T_{lm}(-k^2,t) &= \frac{2}{\pi}\int_0^\infty \cos(\omega t) T_{lm}(-k^2,\omega)d\omega
\end{aligned} \tag{4.27}$$

Here, $\dot{Q}_{lm}(0,\omega)$ and $\dot{Q}_{lm}(-k^2,\omega)$ stand for:

$$\begin{aligned}
\dot{Q}_{lm}(0,\omega) &\equiv \int_0^\infty \cos(\omega t)\dot{Q}_{lm}(0,t)dt \\
\dot{Q}_{lm}(-k^2,\omega) &\equiv \int_0^\infty \cos(\omega t)\dot{Q}_{lm}(-k^2,t)dt \\
\dot{Q}_{lm}(0,t) &= \frac{2}{\pi}\int_0^\infty \cos(\omega t)\dot{Q}_{lm}(0,\omega)d\omega \\
\dot{Q}_{lm}(-k^2,t) &= \frac{2}{\pi}\int_0^\infty \cos(\omega t)\dot{Q}_{lm}(-k^2,\omega)d\omega \\
\dot{Q}_{lm}(0,\omega) &\equiv \omega Q_{lm}(0,\omega) \\
\dot{Q}_{lm}(-\frac{\omega^2}{c^2},\omega) &\equiv \omega Q_{lm}(-\frac{\omega^2}{c^2},\omega)
\end{aligned} \tag{4.28}$$

Now, the continuity relation (2.68) can be rewritten as:

$$\vec{\nabla}\cdot\vec{j}(\vec{r},\omega) + \omega\rho(\vec{r},\omega) = 0 \tag{4.29}$$

Considering these Fourier transforms, one must evaluate the scalar potential:

$$\varphi(\vec{r},t)=\varphi^{(1)}(\vec{r},t)+\varphi^{(2)}(\vec{r},t) \tag{4.30}$$

where

$$\varphi^{(1)}(\vec{r},t)=\left(\frac{2}{\pi}\int_0^\infty \sin(\omega t)d\omega\right)\left(\int \frac{\cos\left(\frac{\omega}{c}|\vec{r}-\vec{r}'|\right)}{|\vec{r}-\vec{r}'|}\rho(\vec{r}',t)d^3\vec{r}'\right) \tag{4.31}$$

$$\varphi^{(2)}(\vec{r},t)=\left(-\frac{2}{\pi}\int_0^\infty \cos(\omega t)d\omega\right)\left(\int \frac{\sin\left(\frac{\omega}{c}|\vec{r}-\vec{r}'|\right)}{|\vec{r}-\vec{r}'|}\rho(\vec{r}',\omega)d^3\vec{r}'\right) \tag{4.32}$$

In addition, the vector potential can be written as:

$$\vec{A}(\vec{r},t)=\vec{A}^{(1)}(\vec{r},t)+\vec{A}^{(2)}(\vec{r},t) \tag{4.33}$$

where

$$\vec{A}^{(1)}(\vec{r},t)=\left(\frac{2}{\pi c}\int_0^\infty \sin(\omega t)d\omega\right)\left(\int \frac{\sin\left(\frac{\omega}{c}|\vec{r}-\vec{r}'|\right)}{|\vec{r}-\vec{r}'|}\vec{j}(\vec{r}',t)d^3\vec{r}'\right) \tag{4.34}$$

$$\vec{A}^{(2)}(\vec{r},t)=\left(\frac{2}{\pi c}\int_0^\infty \cos(\omega t)d\omega\right)\left(\int \frac{\cos\left(\frac{\omega}{c}|\vec{r}-\vec{r}'|\right)}{|\vec{r}-\vec{r}'|}\vec{j}(\vec{r}',\omega)d^3\vec{r}'\right) \tag{4.35}$$

On the other hand, for the scalar potential case:

$$\frac{e^{i(\omega/c)|\vec{r}-\vec{r}'|}}{|\vec{r}-\vec{r}'|}=\frac{\omega}{4\pi c}\sum_{l,m}F^*_{lm(\omega/c)}(\vec{r}')H_{lm(\omega/c)}(\vec{r}),\quad r>r' \tag{4.36}$$

$$\frac{\cos\left(\frac{\omega}{c}|\vec{r}-\vec{r}'|\right)}{|\vec{r}-\vec{r}'|}=-\frac{\omega}{4\pi c}\sum_{l,m}F^*_{lm(\omega/c)}(\vec{r}')N_{lm(\omega/c)}(\vec{r}),\quad r>r' \tag{4.37}$$

$$\frac{\sin\left(\frac{\omega}{c}|\vec{r}-\vec{r}'|\right)}{|\vec{r}-\vec{r}'|}=\frac{\omega}{4\pi c}\sum_{l,m}F^*_{lm(\omega/c)}(\vec{r}')F_{lm(\omega/c)}(\vec{r}),\quad r>r' \tag{4.38}$$

where $F_{lm(\omega/c)}$ was already defined in (2.73), and $H_{lm(\omega/c)}(\vec{r})$ and $N_{lm(\omega/c)}(\vec{r})$ are defined similar to $F_{lm(\omega/c)}$, but with the spherical Bessel function of the first kind j_l replaced with the spherical Hankel function $h_l^{(+)}$ and the spherical Bessel function of the second species n_l, respectively (Radescu and Vaman 2002), as following:

$$H_{lm(\omega/c)}(\vec{r})=h_l^{(+)}\left(\frac{\omega}{c}r\right)Y_{lm}(\vec{n}) \tag{4.39}$$

$$N_{lm(\omega/c)}(\vec{r}) = n_l\left(\frac{\omega}{c}r\right)Y_{lm}(\vec{n}) \tag{4.40}$$

In the equations above, H_{lmk} and N_{lmk} satisfy the same normalization, completeness, and parity conditions as those satisfied by F_{lmk} in (2.75), (2.76), and (2.78):

$$\begin{aligned}\int H_{lmk}(\vec{r})H^*_{l'm'k'}(\vec{r})d^3r &= \int N_{lmk}(\vec{r})N^*_{l'm'k'}(\vec{r})d^3r \\ &= \int F_{lmk}(\vec{r})F^*_{l'm'k'}(\vec{r})d^3r \\ &= \delta_{ll'}\delta_{mm'}\frac{(2\pi)^3}{k^2}\delta(k-k')\end{aligned} \tag{4.41}$$

$$\begin{aligned}\sum_{lmk} H_{lmk}(\vec{r})H^*_{l'm'k'}(\vec{r}') &= \sum_{lmk} N_{lmk}(\vec{r})N^*_{l'm'k'}(\vec{r}') \\ &= \sum_{lmk} N_{lmk}(\vec{r})N^*_{l'm'k'}(\vec{r}') \\ &= (2\pi)^3\delta(\vec{r}-\vec{r}'),\quad \sum_k = \int_0^\infty k^2dk\end{aligned} \tag{4.42}$$

$$H_{lmk}(-\vec{r}) = (-1)^l H_{lmk}(\vec{r}),\ N_{lmk}(-\vec{r}) = (-1)^l N_{lmk}(\vec{r}) \tag{4.43}$$

Besides that, for $\vec{F}^{(\lambda)}_{lmk}$, one should have:

$$\begin{cases}\vec{\nabla}\times\vec{N}^{(-)}_{lmk}(\vec{r}) = 0,\quad \vec{\nabla}\times\vec{H}^{(-)}_{lmk}(\vec{r}) = 0 \\ \vec{\nabla}\cdot\vec{N}^{(-)}_{lmk}(\vec{r}) = ikN_{lmk}(\vec{r}),\quad \vec{\nabla}\cdot\vec{H}^{(-)}_{lmk}(\vec{r}) = ikH_{lmk}(\vec{r}) \\ \vec{\nabla}\cdot\vec{N}^{(+)}_{lmk}(\vec{r}) = \vec{\nabla}\cdot\vec{N}^{(0)}_{lmk}(\vec{r}) = 0 \\ \vec{\nabla}\cdot\vec{H}^{(+)}_{lmk}(\vec{r}) = \vec{\nabla}\cdot\vec{H}^{(0)}_{lmk}(\vec{r}) = 0\end{cases} \tag{4.44}$$

For the vector potential, it would be helpful to have the representations for the Green's functions in terms of the basis vector functions for the vector Helmholtz equation:

$$\frac{e^{i(\omega/c)|\vec{r}-\vec{r}'|}}{|\vec{r}-\vec{r}'|} = \frac{\omega}{4\pi c}\sum_{l,m}\vec{F}^{(\lambda)*}_{lm(\omega/c)}(\vec{r}')\cdot\vec{H}^{(\lambda)}_{lm(\omega/c)}(\vec{r}),\quad r>r' \tag{4.45}$$

which results from the slightly more general expansion:

$$\frac{1}{3}\delta_{ik}\frac{e^{i(\omega/c)|\vec{r}-\vec{r}'|}}{|\vec{r}-\vec{r}'|} = \frac{\omega}{4\pi c}\sum_{l,m,\lambda}\left[\vec{F}^{(\lambda)*}_{lm(\omega/c)}(\vec{r}')\right]_i\cdot\left[\vec{H}^{(\lambda)}_{lm(\omega/c)}(\vec{r})\right]_k,\quad r>r' \tag{4.46}$$

$$\frac{\cos\left(\frac{\omega}{c}|\vec{r}-\vec{r}'|\right)}{|\vec{r}-\vec{r}'|} = -\frac{\omega}{4\pi c}\sum_{l,m}\Big[\vec{F}^{(+)*}_{lm(\omega/c)}(\vec{r}')\cdot\vec{N}^{(+)}_{lm(\omega/c)}(\vec{r}) + \vec{F}^{(-)*}_{lm(\omega/c)}(\vec{r}')\cdot\vec{N}^{(-)}_{lm(\omega/c)}(\vec{r}) + \vec{F}^{(0)*}_{lm(\omega/c)}(\vec{r}')\cdot\vec{N}^{(0)}_{lm(\omega/c)}(\vec{r})\Big],\quad r>r' \tag{4.47}$$

$$\frac{\sin\left(\frac{\omega}{c}|\vec{r}-\vec{r}'|\right)}{|\vec{r}-\vec{r}'|} = \frac{\omega}{4\pi c}\sum_{l,m}\Big[\vec{F}^{(+)*}_{lm(\omega/c)}(\vec{r}')\cdot\vec{F}^{(+)}_{lm(\omega/c)}(\vec{r}) \; \vec{F}^{(-)*}_{lm(\omega/c)}(\vec{r}')\cdot\vec{F}^{(-)}_{lm(\omega/c)}(\vec{r}) + \vec{F}^{(0)*}_{lm(\omega/c)}(\vec{r}')\cdot\vec{F}^{(0)}_{lm(\omega/c)}(\vec{r})\Big],\quad r>r' \tag{4.48}$$

where $\vec{H}^{(\lambda)}_{lmk}$ and $\vec{N}^{(\lambda)}_{lmk}$ $(\lambda=0,\pm)$ are defined similar to $\vec{F}^{(\lambda=0,\pm)}_{lmk}$, but with $h^{(+)}_l$ and n_l instead of j_l:

$$\begin{pmatrix}\vec{H}^{(0)}_{lmk}(\vec{r})\\ \vec{N}^{(0)}_{lmk}(\vec{r})\end{pmatrix} = \frac{i}{\sqrt{l(l+1)}}\vec{\nabla}\times\begin{pmatrix}\vec{r}H_{lmk}(\vec{r})\\ \vec{r}N_{lmk}(\vec{r})\end{pmatrix} = \begin{pmatrix}h^{(+)}_l(kr)\\ n_l(kr)\end{pmatrix}\vec{Y}_{llm}(\vec{n}) \tag{4.49}$$

$$\begin{pmatrix}\vec{H}^{(+)}_{lmk}(\vec{r})\\ \vec{N}^{(+)}_{lmk}(\vec{r})\end{pmatrix} = -\frac{1}{\sqrt{l(l+1)}}\frac{i}{k}\vec{\nabla}\times\vec{\nabla}\times\begin{pmatrix}\vec{r}H_{lmk}(\vec{r})\\ \vec{r}N_{lmk}(\vec{r})\end{pmatrix} \tag{4.50}$$

$$\begin{pmatrix}\vec{H}^{(-)}_{lmk}(\vec{r})\\ \vec{N}^{(-)}_{lmk}(\vec{r})\end{pmatrix} = -\frac{i}{k}\vec{\nabla}\begin{pmatrix}\vec{r}H_{lmk}(\vec{r})\\ \vec{r}N_{lmk}(\vec{r})\end{pmatrix} \tag{4.51}$$

and

$$\begin{aligned}\int \vec{H}^{(\lambda)*}_{lmk}(\vec{r})\vec{H}^{(\lambda')}_{l'm'k'}(\vec{r})d^3\vec{r} &= \int \vec{N}^{(\lambda)*}_{lmk}(\vec{r})\vec{N}^{(\lambda')}_{l'm'k'}(\vec{r})d^3\vec{r}\\ &= \int \vec{F}^{(\lambda)*}_{lmk}(\vec{r})\vec{F}^{(\lambda')}_{l'm'k'}(\vec{r})d^3\vec{r}\\ &= \delta_{ll'}\delta_{mm'}\delta_{\lambda\lambda'}\frac{(2\pi)^3}{k^2}\delta(k-k')\end{aligned} \tag{4.52}$$

$$\begin{aligned}\sum_{l,m,k,\lambda}\Big[\vec{H}^{(\lambda)}_{lmk}(\vec{r})\Big]^*_i\Big[\vec{H}^{(\lambda)}_{lmk}(\vec{r}')\Big]_j &= \sum_{l,m,k,\lambda}\Big[\vec{N}^{(\lambda)}_{lmk}(\vec{r})\Big]^*_i\Big[\vec{N}^{(\lambda)}_{lmk}(\vec{r}')\Big]_j\\ &= (2\pi)^3\delta_{ij}\delta(\vec{r}-\vec{r}')\end{aligned} \tag{4.53}$$

$$\vec{H}^{(\lambda)}_{lmk}(-\vec{r}) = (-1)^{l+\lambda}\vec{H}^{(\lambda)}_{lmk}(\vec{r}) \tag{4.54}$$

$$\vec{N}^{(\lambda)}_{lmk}(-\vec{r}) = (-1)^{l+\lambda}\vec{N}^{(\lambda)}_{lmk}(\vec{r}),\ \lambda = 0,\pm \tag{4.55}$$

Here, (4.47) and (4.48) can be derived straightforwardly through separating the real and imaginary parts in (4.46).

Utilizing the form factors presented above for the Green's functions, the scalar potential can be obtained from (4.30) to (4.32) in the following form:

$$\varphi(\vec{r},t)=\varphi^{(1)}(\vec{r},t)+\varphi^{(2)}(\vec{r},t)$$

$$\varphi^{(1)}(\vec{r},t)=-\frac{1}{\pi^{3/2}}\int_0^\infty \omega\sin(\omega t)d\omega\left(\sum_{l,m}\left(\frac{-i\omega}{c}\right)^l\frac{\sqrt{2l+1}}{(2l+1)!!}\right)$$
$$\times\, Q_{lm}\left(-\frac{\omega^2}{c^2},\omega\right)N_{lm(\omega/c)}(\vec{r})$$

$$\varphi^{(2)}(\vec{r},t)=-\frac{1}{\pi^{3/2}}\int_0^\infty \omega\cos(\omega t)d\omega\left(\sum_{l,m}\left(\frac{-i\omega}{c}\right)^l\frac{\sqrt{2l+1}}{(2l+1)!!}\right)$$
$$\times\, Q_{lm}\left(-\frac{\omega^2}{c^2},\omega\right)F_{lm(\omega/c)}(\vec{r}) \tag{4.56}$$

Indeed, the equation above can be rewritten in much simpler and compact format:

$$\varphi(\vec{r},t)=-\frac{1}{\pi^{3/2}}\int_0^\infty \omega d\omega\left(\sum_{lm}\left(\frac{-i\omega}{c}\right)^l\frac{\sqrt{2l+1}}{(2l+1)!!}\left[\sin(\omega t)N_{lm(\omega/c)}(\vec{r})\right.\right.$$
$$\left.+\cos(\omega t)F_{lm(\omega/c)}(\vec{r})\right]Q_{lm}\left(-\frac{\omega^2}{c^2},\omega\right) \tag{4.57}$$

Similar approach can also be conducted for the vector potential equations in (4.33)–(4.35):

$$\vec{A}(\vec{r},t)=\vec{A}^{(1)}(\vec{r},t)+\vec{A}^{(2)}(\vec{r},t)$$

$$\vec{A}^{(1)}(\vec{r},t)=-\frac{1}{2\pi^2c^2}\int_0^\infty \omega\cos(\omega t)d\omega\left(\sum_{l,m}\left(\frac{-i\omega}{c}\right)^{l-1}\frac{\sqrt{4\pi(2l+1)(l+1)}}{\sqrt{l}(2l+1)!!}\right)$$
$$\times\left\{\omega M_{lm}\left(-\frac{\omega^2}{c^2},\omega\right)\vec{N}^{(0)}_{lm(\omega/c)}(\vec{r})+\left[\omega Q_{lm}(0,\omega)\right.\right.$$
$$\left.+\frac{\omega^2}{c}T_{lm}\left(-\frac{\omega^2}{c^2},\omega\right)\right]\vec{N}^{(+)}_{lm(\omega/c)}(\vec{r})$$
$$\left.+\frac{\sqrt{l}}{\sqrt{l+1}}\omega Q_{lm}\left(-\frac{\omega^2}{c^2},\omega\right)\vec{N}^{(-)}_{lm(\omega/c)}(\vec{r})\right\}$$

$$\vec{A}^{(2)}(\vec{r},t) = \frac{1}{2\pi^2 c^2}\int\limits_0^\infty \omega \sin(\omega t) d\omega \left(\sum_{l,m} \left(\frac{-i\omega}{c}\right)^{l-1} \frac{\sqrt{4\pi(2l+1)(l+1)}}{\sqrt{l}(2l+1)!!} \right)$$
$$\times \left\{ \omega M_{lm}\left(-\frac{\omega^2}{c^2},\omega\right) \vec{F}^{(0)}_{lm(\omega/c)}(\vec{r}) + [\omega Q_{lm}(0,\omega)\right.$$
$$\left. + \frac{\omega^2}{c} T_{lm}\left(-\frac{\omega^2}{c^2},\omega\right)\right] \vec{F}^{(+)}_{lm(\omega/c)}(\vec{r}) + \frac{\sqrt{l}}{\sqrt{l+1}} \omega Q_{lm}\left(-\frac{\omega^2}{c^2},\omega\right) \vec{F}^{(-)}_{lm(\omega/c)}(\vec{r}) \Big\} \tag{4.58}$$

Thus, with (4.21), (4.56), and (4.58):

$$\begin{aligned} \vec{\nabla} F_{lm(\omega/c)}(\vec{r}) &= i\frac{\omega}{c} F^{(-)}_{lm(\omega/c)}(\vec{r}) \\ \vec{\nabla} N_{lm(\omega/c)}(\vec{r}) &= i\frac{\omega}{c} \vec{N}^{(-)}_{lm(\omega/c)}(\vec{r}) \end{aligned} \tag{4.59}$$

The findings above lead us to the expression for the electric field $\vec{E}(\vec{r},t)$ emitted by the most general type of source described by the electric, magnetic, and toroidal multipole form factors, as previously discussed in (2.88). The analogous description for the magnetic field was presented in (2.89), which is obtained through taking the curls of $\vec{N}^{(\lambda)}_{lm(\omega/c)}(\vec{r})$ and $\vec{F}^{(\lambda)}_{lm(\omega/c)}(\vec{r})$ using the following:

$$\begin{cases} \vec{\nabla} \times \vec{N}^{(-)}_{lmk}(\vec{r}) = \vec{\nabla} \times \vec{F}^{(-)}_{lmk}(\vec{r}) = 0 \\ \vec{\nabla} \times \vec{F}^{(0)}_{lmk}(\vec{r}) = -\frac{\omega}{c} \vec{F}^{(+)}_{lmk}(\vec{r}),\ \vec{\nabla} \times \vec{F}^{(+)}_{lmk}(\vec{r}) = -\frac{\omega}{c} \vec{F}^{(0)}_{lmk}(\vec{r}) \\ \vec{\nabla} \times \vec{N}^{(0)}_{lmk}(\vec{r}) = -\frac{\omega}{c} \vec{N}^{(+)}_{lmk}(\vec{r}),\ \vec{\nabla} \times \vec{N}^{(+)}_{lmk}(\vec{r}) = -\frac{\omega}{c} \vec{N}^{(0)}_{lmk}(\vec{r}) \end{cases} \tag{4.60}$$

The equations for the electric (2.88) and magnetic (2.89) fields express that the multipole content of the source reflects itself in the fields that are created based on electric, magnetic, and toroidal multipole form factors. The descriptions for both $\vec{E}(\vec{r},t)$ and $\vec{B}(\vec{r},t)$ were obtained from the retarded scalar and vector potentials $\varphi(\vec{r},t)$ and $\vec{A}(\vec{r},t)$, respectively, and expressed in terms of the electric, magnetic, and toroidal multipole components as in (4.56). Although the equations are clear enough, some remarks on these equations may still be required in connection with gauge invariance. Besides, the elucidated potentials $\varphi(\vec{r},t)$ and $\vec{A}(\vec{r},t)$ satisfy the Lorenz condition as:

$$\vec{\nabla} \cdot \vec{A}(\vec{r},t) + \frac{1}{c}\frac{\partial \varphi(\vec{r},t)}{\partial t} = 0 \tag{4.61}$$

To check this condition, the following procedure can be applied:

$$\begin{cases} \vec{\nabla}\cdot\vec{N}^{(0)}_{lmk}(\vec{r}) = 0 \\ \vec{\nabla}\cdot\vec{N}^{(+)}_{lmk}(\vec{r}) = 0 \\ \vec{\nabla}\cdot\vec{N}^{(-)}_{lmk}(\vec{r}) = ikN_{lmk}(\vec{r}) \end{cases} \tag{4.62}$$

where $k = \omega/c$. Then:

$$\begin{aligned} \vec{\nabla}\cdot\vec{A}^{(1)}(\vec{r},t) &= \frac{1}{2\pi^2c^2}\int_0^\infty \omega^2\cos(\omega t)d\omega\left(\sum_{l,m}\left(\frac{-i\omega}{c}\right)^l\frac{\sqrt{4\pi(2l+1)}}{(2l+1)!!}\right. \\ &\quad\left.\times\, Q_{lm}\left(-\frac{\omega^2}{c^2},\omega\right)N_{lm(\omega/c)}(\vec{r})\right) \\ \vec{\nabla}\cdot\vec{A}^{(2)}(\vec{r},t) &= -\frac{1}{2\pi^2c^2}\int_0^\infty \omega^2\sin(\omega t)d\omega\left(\sum_{l,m}\left(\frac{-i\omega}{c}\right)^l\frac{\sqrt{4\pi(2l+1)}}{(2l+1)!!}\right. \\ &\quad\left.\times\, Q_{lm}\left(-\frac{\omega^2}{c^2},\omega\right)F_{lm(\omega/c)}(\vec{r})\right) \end{aligned} \tag{4.63}$$

and

$$\begin{aligned} \frac{1}{c}\frac{\partial\varphi^{(1)}(\vec{r},t)}{\partial t} &= -\frac{1}{\pi^{3/2}c^2}\int_0^\infty \omega^2\cos(\omega t)d\omega\left(\sum_{l,m}\left(\frac{-i\omega}{c}\right)^l\frac{\sqrt{(2l+1)}}{(2l+1)!!}\right. \\ &\quad\left.\times\, Q_{lm}\left(-\frac{\omega^2}{c^2},\omega\right)N_{lm(\omega/c)}(\vec{r})\right) \\ \frac{1}{c}\frac{\partial\varphi^{(2)}(\vec{r},t)}{\partial t} &= \frac{1}{\pi^{3/2}c^2}\int_0^\infty \omega^2\sin(\omega t)d\omega\left(\sum_{l,m}\left(\frac{-i\omega}{c}\right)^l\frac{\sqrt{(2l+1)}}{(2l+1)!!}\right. \\ &\quad\left.\times\, Q_{lm}\left(-\frac{\omega^2}{c^2},\omega\right)F_{lm(\omega/c)}(\vec{r})\right) \end{aligned} \tag{4.64}$$

Therefore, it is possible to consider these components separately as:

$$\vec{\nabla}\cdot\vec{A}^{(1)}(\vec{r},t) + \frac{1}{c}\frac{\partial\varphi^{(1)}(\vec{r},t)}{\partial t} = 0 \tag{4.65}$$

$$\vec{\nabla}\cdot\vec{A}^{(2)}(\vec{r},t) + \frac{1}{c}\frac{\partial\varphi^{(2)}(\vec{r},t)}{\partial t} = 0 \tag{4.66}$$

and the Lorenz condition in (4.61) for $\vec{A}(\vec{r},t) = \vec{A}^{(1)}(\vec{r},t) + \vec{A}^{(2)}(\vec{r},t)$ and $\varphi(\vec{r},t) = \varphi^{(1)}(\vec{r},t) + \varphi^{(2)}(\vec{r},t)$ is verified.

Here, Lorenz gauge was utilized to express $\vec{A}$ and φ. To demonstrate that the gauge invariance still valid after satisfying the Lorenz condition, a more general form of the potentials is required:

$$\varphi'(\vec{r},t)=\varphi'^{(1)}(\vec{r},t)+\varphi'^{(2)}(\vec{r},t)$$

$$\varphi'^{(1)}(\vec{r},t)=-\frac{C_1}{\pi^{3/2}c}\int\limits_0^\infty \omega\sin(\omega t)d\omega\left(\sum_{l,m}\left(\frac{-i\omega}{c}\right)^l\frac{\sqrt{2l+1}}{(2l+1)!!}\right.$$
$$\left.\times\, Q_{lm}\left(-\frac{\omega^2}{c^2},\omega\right)N_{lm(\omega/c)}(\vec{r})\right)$$

$$\varphi^{(2)}(\vec{r},t)=-\frac{C_2}{\pi^{3/2}c}\int\limits_0^\infty \omega\cos(\omega t)d\omega\left(\sum_{l,m}\left(\frac{-i\omega}{c}\right)^l\frac{\sqrt{2l+1}}{(2l+1)!!}\right.$$
$$\left.\times\, Q_{lm}\left(-\frac{\omega^2}{c^2},\omega\right)F_{lm(\omega/c)}(\vec{r})\right) \tag{4.67}$$

$$\vec{A}'(\vec{r},t)=\vec{A}'^{(1)}(\vec{r},t)+\vec{A}'^{(2)}(\vec{r},t)$$

$$\vec{A}'^{(1)}(\vec{r},t)=-\frac{1}{2\pi^2c^2}\int\limits_0^\infty \omega\cos(\omega t)d\omega\left(\sum_{l,m}\left(\frac{-i\omega}{c}\right)^{l-1}\frac{\sqrt{4\pi(2l+1)(l+1)}}{\sqrt{l}(2l+1)!!}\right.$$
$$\times\left\{\omega M_{lm}\left(-\frac{\omega^2}{c^2},\omega\right)\vec{N}^{(0)}_{lm(\omega/c)}(\vec{r})+\left[\omega Q_{lm}(0,\omega)\right.\right.$$
$$\left.+\frac{\omega^2}{c}T_{lm}\left(-\frac{\omega^2}{c^2},\omega\right)\right]\vec{N}^{(+)}_{lm(\omega/c)}(\vec{r})$$
$$\left.\left.+\frac{C_1\sqrt{l}}{\sqrt{l+1}}\omega Q_{lm}\left(-\frac{\omega^2}{c^2},\omega\right)\vec{N}^{(-)}_{lm(\omega/c)}(\vec{r})\right\}\right)$$

$$\vec{A}^{(2)}(\vec{r},t)=\frac{1}{2\pi^2c^2}\int\limits_0^\infty \omega\sin(\omega t)d\omega\left(\sum_{l,m}\left(\frac{-i\omega}{c}\right)^{l-1}\frac{\sqrt{4\pi(2l+1)(l+1)}}{\sqrt{l}(2l+1)!!}\right.$$
$$\times\left\{\omega M_{lm}\left(-\frac{\omega^2}{c^2},\omega\right)\vec{F}^{(0)}_{lm(\omega/c)}(\vec{r})+\left[\omega Q_{lm}(0,\omega)\right.\right.$$
$$\left.\left.+\frac{\omega^2}{c}T_{lm}\left(-\frac{\omega^2}{c^2},\omega\right)\right]\vec{F}^{(+)}_{lm(\omega/c)}(\vec{r})+\frac{C_2\sqrt{l}}{\sqrt{l+1}}\omega Q_{lm}\left(-\frac{\omega^2}{c^2},\omega\right)\vec{F}^{(-)}_{lm(\omega/c)}(\vec{r})\right\}\right) \tag{4.68}$$

The new potentials $\varphi'(\vec{r},t)$ and $\vec{A}'(\vec{r},t)$ in (4.67) and (4.68), respectively, and their two components, separately specified by the superscripts (1) and (2): $\varphi^{(1),(2)}(\vec{r},t)$ and $\vec{A}^{(1),(2)}(\vec{r},t)$, are related to the previous ones through the following gauge transformation:

$$\varphi'^{(1)}(\vec{r},t) = \varphi^{(1)}(\vec{r},t) - \frac{1}{c}\frac{\partial \Lambda^{(1)}(\vec{r},t)}{\partial t}$$
$$\vec{A}'^{(1)}(\vec{r},t) = \vec{A}^{(1)}(\vec{r},t) + \vec{\nabla}\Lambda^{(1)}(\vec{r},t) \tag{4.69}$$

$$\varphi'^{(2)}(\vec{r},t) = \varphi^{(2)}(\vec{r},t) - \frac{1}{c}\frac{\partial \Lambda^{(2)}(\vec{r},t)}{\partial t}$$
$$\vec{A}'^{(2)}(\vec{r},t) = \vec{A}^{(2)}(\vec{r},t) + \vec{\nabla}\Lambda^{(2)}(\vec{r},t) \tag{4.70}$$

$$\varphi'(\vec{r},t) = \varphi(\vec{r},t) - \frac{1}{c}\frac{\partial \Lambda(\vec{r},t)}{\partial t}, \quad \vec{A}'(\vec{r},t) = \vec{A}(\vec{r},t) + \vec{\nabla}\Lambda(\vec{r},t) \tag{4.71}$$

where

$$\Lambda(\vec{r},t) = \Lambda^{(1)}(\vec{r},t) + \Lambda^{(2)}(\vec{r},t) \tag{4.72}$$

$$\Lambda^{(1)}(\vec{r},t) \equiv \frac{1}{2\pi^2c^2}\int_0^\infty \omega\cos(\omega t)d\omega\left(\sum_{l,m}\left(\frac{-i\omega}{c}\right)^{l-1}\frac{\sqrt{4\pi(2l+1)}}{(2l+1)!!}\right)$$
$$\times\left\{\omega Q_{lm}\left(-\frac{\omega^2}{c^2},\omega\right)\frac{ic}{\omega}(C_1-1)N_{lm(\omega/c)}(\vec{r})\right\} \tag{4.73}$$

$$\Lambda^{(2)}(\vec{r},t) \equiv \frac{1}{2\pi^2c^2}\int_0^\infty \omega\cos(\omega t)d\omega\left(\sum_{l,m}\left(\frac{-i\omega}{c}\right)^{l-1}\frac{\sqrt{4\pi(2l+1)}}{(2l+1)!!}\right)$$
$$\times\left\{\omega Q_{lm}\left(-\frac{\omega^2}{c^2},\omega\right)\frac{ic}{\omega}(C_2-1)F_{lm(\omega/c)}(\vec{r})\right\} \tag{4.74}$$

Now, (4.69)–(4.72) could be assessed by noting that:

$$\vec{N}^{(-)}_{lm(\omega/c)}(\vec{r}) = -\frac{ic}{\omega}\vec{\nabla}N_{lm(\omega/c)}(\vec{r}) \tag{4.75}$$

$$\vec{F}^{(-)}_{lm(\omega/c)}(\vec{r}) = -\frac{ic}{\omega}\vec{\nabla}F_{lm(\omega/c)}(\vec{r}) \tag{4.76}$$

Since

$$\left(\Delta+k^2\right)N_{lm(\omega/c)}(\vec{r}) = 0, \quad \left(\Delta+k^2\right)F_{lm(\omega/c)}(\vec{r}) = 0 \tag{4.77}$$

the gauge functions $\Lambda^{(1)}$, $\Lambda^{(2)}$, and $\Lambda = \Lambda^{(1)} + \Lambda^{(2)}$ satisfy the wave equation:

$$\left(\Delta - \frac{1}{c^2}\frac{\partial^2}{\partial t^2}\right)\Lambda^{(1)}(\vec{r},t) = 0 \tag{4.78}$$

$$\left(\Delta - \frac{1}{c^2}\frac{\partial^2}{\partial t^2}\right)\Lambda^{(2)}(\vec{r}, t) = 0 \tag{4.79}$$

$$\left(\Delta - \frac{1}{c^2}\frac{\partial^2}{\partial t^2}\right)\Lambda(\vec{r}, t) = 0 \tag{4.80}$$

In fact, the new potentials $\vec{A}'(\vec{r}, t)$ and $\varphi'(\vec{r}, t)$ from (4.67) to (4.68) still satisfy the Lorenz condition:

$$\vec{\nabla} \cdot \vec{A}'^{(1)}(\vec{r}, t) + \frac{1}{c}\frac{\partial \varphi'^{(1)}(\vec{r}, t)}{\partial t} = 0 \tag{4.81}$$

$$\vec{\nabla} \cdot \vec{A}'^{(2)}(\vec{r}, t) + \frac{1}{c}\frac{\partial \varphi'^{(2)}(\vec{r}, t)}{\partial t} = 0 \tag{4.82}$$

$$\vec{\nabla} \cdot \vec{A}'(\vec{r}, t) + \frac{1}{c}\frac{\partial \varphi'(\vec{r}, t)}{\partial t} = 0 \tag{4.83}$$

Overall, the new potentials $\varphi'(\vec{r}, t)$ and $\vec{A}'(\vec{r}, t)$ in (4.67) and (4.68), respectively, are more general than the previous versions in (4.56) and (4.58) and are gauge equivalent to them. Although both forms of the potentials satisfy the Lorenz condition, in the new ones, the remaining gauge invariance is illustrated through the two remaining real arbitrary constants C_1 and C_2. Here, the previous potentials can be obtained as particular cases from the new ones when:

$$C_1 = C_2 = C = 1 \tag{4.84}$$

The condition above fixes the gauge in (4.56) and (4.58). Another convenient gauge that one can use is Coulomb gauge, by defining:

$$C_1 = C_2 = C = 0 \tag{4.85}$$

when the potentials φ'' and $\vec{A}''$ satisfy the following conditions:

$$\varphi''(\vec{r}, t) \equiv 0, \quad \nabla \vec{A}''(\vec{r}, t) \equiv 0 \tag{4.86}$$

Because of their gauge uniformity that was argued before, all these forms of the potentials give rise to the identical fields $\vec{E}(\vec{r}, t)$ and $\vec{B}(\vec{r}, t)$, and since these fields are considered here, all the results are gauge invariant.

4.1.2 Fields at Far Distances

Obtaining the expressions for the electric (2.88) and magnetic (2.89) fields at far distances allows to derive the radiation intensity, angular momentum loss, and recoil force. To this end, we will consider the behavior of the fields $\vec{E}(\vec{r},t)$ and $\vec{B}(\vec{r},t)$ to the order $O(1/r)$ (representing large distances), and the asymptotical behavior of the spherical Bessel functions $j_l(\omega r/c)$ and $n_l(\omega r/c)$, for $r \to \infty$:

$$j_l\left(\frac{\omega}{c}r\right) \sim 4\pi i^l \frac{\sin\left(\left(\frac{\omega}{c}r\right) - l\frac{\pi}{2}\right)}{\left(\frac{\omega}{c}r\right)} \tag{4.87}$$

$$n_l\left(\frac{\omega}{c}r\right) \sim -4\pi i^l \frac{\cos\left(\left(\frac{\omega}{c}r\right) - l\frac{\pi}{2}\right)}{\left(\frac{\omega}{c}r\right)} \tag{4.88}$$

$$j_{l+1}\left(\frac{\omega}{c}r\right) \sim -4\pi i^{l+1} \frac{\cos\left(\left(\frac{\omega}{c}r\right) - l\frac{\pi}{2}\right)}{\left(\frac{\omega}{c}r\right)} \tag{4.89}$$

$$n_{l+1}\left(\frac{\omega}{c}r\right) \sim -4\pi i^{l+1} \frac{\sin\left(\left(\frac{\omega}{c}r\right) - l\frac{\pi}{2}\right)}{\left(\frac{\omega}{c}r\right)} \tag{4.90}$$

$$j_{l-1}\left(\frac{\omega}{c}r\right) \sim 4\pi i^{l-1} \frac{\cos\left(\left(\frac{\omega}{c}r\right) - l\frac{\pi}{2}\right)}{\left(\frac{\omega}{c}r\right)} \tag{4.91}$$

$$n_{l-1}\left(\frac{\omega}{c}r\right) \sim 4\pi i^{l-1} \frac{\sin\left(\left(\frac{\omega}{c}r\right) - l\frac{\pi}{2}\right)}{\left(\frac{\omega}{c}r\right)} \tag{4.92}$$

To obtain the $O(1/r)$ definition of the vector functions $\vec{F}^{(\lambda)}_{lm(\omega/c)}(\vec{r})$ and $\vec{N}^{(\lambda)}_{lm(\omega/c)}(\vec{r})$, one should consider the followings:

$$\vec{F}^{(0)}_{lm(\omega/c)}(\vec{r}) \overset{O(1/r)}{\sim} \frac{4\pi i^l}{\sqrt{2l+1}} \frac{\sin\left(\left(\frac{\omega}{c}r\right) - l\frac{\pi}{2}\right)}{\left(\frac{\omega}{c}r\right)} \vec{Y}_{llm}(\vec{n}) \tag{4.93}$$

$$\vec{F}^{(+)}_{lm(\omega/c)}(\vec{r}) \overset{O(1/r)}{\sim} \frac{4\pi i^{l-1}}{\sqrt{2l+1}} \frac{\cos\left(\left(\frac{\omega}{c}r\right) - l\frac{\pi}{2}\right)}{\left(\frac{\omega}{c}r\right)} \left[\sqrt{l}\vec{Y}_{ll+1m}(\vec{n}) + \sqrt{l+1}\vec{Y}_{ll-1m}(\vec{n})\right] \tag{4.94}$$

$$\vec{F}^{(-)}_{lm(\omega/c)}(\vec{r}) \overset{O(1/r)}{\sim} \frac{4\pi i^{l-1}}{\sqrt{2l+1}} \frac{\cos\left(\left(\frac{\omega}{c}r\right) - l\frac{\pi}{2}\right)}{\left(\frac{\omega}{c}r\right)} \left[\sqrt{l}\vec{Y}_{ll-1m}(\vec{n}) - \sqrt{l+1}\vec{Y}_{ll+1m}(\vec{n})\right] \tag{4.95}$$

$$\vec{N}^{(0)}_{lm(\omega/c)}(\vec{r}) \overset{O(1/r)}{\sim} -\frac{4\pi i^l}{\sqrt{2l+1}} \frac{\cos\left(\left(\frac{\omega}{c}r\right) - l\frac{\pi}{2}\right)}{\left(\frac{\omega}{c}r\right)} \vec{Y}_{llm}(\vec{n}) \tag{4.96}$$

$$\vec{N}^{(+)}_{lm(\omega/c)}(\vec{r}) \overset{O(1/r)}{\sim} \frac{4\pi i^{l-1}}{\sqrt{2l+1}} \frac{\sin\left(\left(\frac{\omega}{c}r\right) - l\frac{\pi}{2}\right)}{\left(\frac{\omega}{c}r\right)} \left[\sqrt{l}\vec{Y}_{ll+1m}(\vec{n}) + \sqrt{l+1}\vec{Y}_{ll-1m}(\vec{n})\right] \tag{4.97}$$

$$\vec{N}^{(-)}_{lm(\omega/c)}(\vec{r}) \overset{O(1/r)}{\sim} \frac{4\pi i^{l-1}}{\sqrt{2l+1}} \frac{\sin\left(\left(\frac{\omega}{c}r\right) - l\frac{\pi}{2}\right)}{\left(\frac{\omega}{c}r\right)} \left[\sqrt{l}\vec{Y}_{ll-1m}(\vec{n}) - \sqrt{l+1}\vec{Y}_{ll+1m}(\vec{n})\right] \tag{4.98}$$

where $\vec{n} = \vec{r}/r$. Therefore, $\vec{E}(\vec{r},t)$ and $\vec{B}(\vec{r},t)$ can be written in terms of the multipole form factors $M_{lm}(-\omega^2/c^2, \omega)$ and $T_{lm}(-\omega^2/c^2, \omega)$, and the electric multipole moments $Q_{lm}(0, \omega)$:

$$\begin{aligned}\vec{E}(\vec{r},t) \overset{O(1/r)}{\sim} & -\frac{2}{\pi c r}\int_0^\infty d\omega \sum_{l,m} \frac{\omega^{l+1}}{c^l} \frac{\sqrt{4\pi(2l+1)(l+1)}}{\sqrt{l}(2l+1)!!} \left\{ -iM_{lm}\left(-\frac{\omega^2}{c^2}, \omega\right) \right. \\ & \times \sin\left(\omega t - \frac{\omega}{c}r + l\frac{\pi}{2}\right)\vec{Y}_{llm}(\vec{n}) + \left[Q_{lm}(0,\omega) + \frac{\omega}{c}T_{lm}\left(-\frac{\omega^2}{c^2}, \omega\right)\right] \\ & \times \cos\left(\omega t - \frac{\omega}{c}r + l\frac{\pi}{2}\right) \\ & \left. \times \left[\frac{\sqrt{l}}{\sqrt{2l+1}}\vec{Y}_{ll+1m}(\vec{n}) + \frac{\sqrt{l+1}}{\sqrt{2l+1}}\vec{Y}_{ll-1m}(\vec{n})\right]\right\}\end{aligned} \tag{4.99}$$

$$\begin{aligned}\vec{B}(\vec{r},t) \overset{O(1/r)}{\sim} & -\frac{2}{\pi r}\int_0^\infty d\omega \sum_{l,m} \frac{\omega^{l+1}}{c^{l+1}} \frac{\sqrt{4\pi(2l+1)(l+1)}}{\sqrt{l}(2l+1)!!} \left\{ iM_{lm}\left(-\frac{\omega^2}{c^2}, \omega\right) \right. \\ & \times \sin\left(\omega t - \frac{\omega}{c}r + l\frac{\pi}{2}\right)\left[\frac{\sqrt{l}}{\sqrt{2l+1}}\vec{Y}_{ll+1m}(\vec{n}) + \frac{\sqrt{l+1}}{\sqrt{2l+1}}\vec{Y}_{ll-1m}(\vec{n})\right] \\ & + \left[Q_{lm}(0,\omega) + \frac{\omega}{c}T_{lm}\left(-\frac{\omega^2}{c^2}, \omega\right)\right] \\ & \left. \times \cos\left(\omega t - \frac{\omega}{c}r + l\frac{\pi}{2}\right)\vec{Y}_{llm}(\vec{n})\right\}\end{aligned} \tag{4.100}$$

The obtained electric and magnetic fields in (4.99) and (4.100), respectively, confirm the transversality condition in the wave zone:

$$\frac{\vec{r}}{r} \times \vec{E}(\vec{r},t) = \vec{B}(\vec{r},t) \tag{4.101}$$

Introducing double-superscript quantities to (4.13)–(4.18) allows us to simplify the equations by removing the integral over ω. These double-superscripts are indeed time derivatives of the multipole mean-square radii of any type (electric, magnetic, toroid) and order (marked by the first superscript). Here, the magnetic field $\vec{B}(\vec{r},t)$ in (4.99) will be treated based on separate contributions from the terms with $l = even$

and $l = odd$ to the sum over l. In the $l = even$ regime, $l = 2\,k$ (where k is an integer), with the Fourier transformations of (4.24), (4.26), and (4.27), one can write:

$$\omega^{l+1} M_{lm}\left(-\frac{\omega^2}{c^2}, \omega\right) = (-1)^{l/2+1} \int_0^\infty dt' M_{ml}^{(l+1)}\left(-\frac{\omega^2}{c^2}, t'\right) \sin(\omega t') \tag{4.102}$$

$$\omega^{l+1} Q_{lm}(0, \omega) = (-1)^{l/2} \int_0^\infty dt' Q_{ml}^{(l+1)}(0, t') \cos(\omega t') \tag{4.103}$$

$$\omega^{l+2} T_{lm}\left(-\frac{\omega^2}{c^2}, \omega\right) = (-1)^{l/2+1} \int_0^\infty dt' T_{ml}^{(l+2)}\left(-\frac{\omega^2}{c^2}, t'\right) \cos(\omega t') \tag{4.104}$$

where the (single) superscript expresses the order of derivation related to the second (time) argument of the form factors. Utilizing:

$$\sin\left(\omega t - \frac{\omega r}{c} + l\frac{\pi}{2}\right) = (-1)^{l/2} \sin\left(\omega t - \frac{\omega r}{c}\right) \tag{4.105}$$

$$\cos\left(\omega t - \frac{\omega r}{c} + l\frac{\pi}{2}\right) = (-1)^{l/2} \cos\left(\omega t - \frac{\omega r}{c}\right) \tag{4.106}$$

in (4.100) and subsequently in (4.102), one can obtain the contribution to the magnetic field from the terms involving:

$$\int_0^\infty d\omega \omega^{l+1} M_{lm}\left(-\frac{\omega^2}{c^2}, \omega\right) \sin\left(\omega t - \frac{\omega r}{c}\right)$$
$$= \frac{(-1)^{l/2+1}}{2} \int_0^\infty d\omega \int_0^\infty dt' M_{lm}^{(l+1)}\left(-\frac{\omega^2}{c^2}, t'\right) \cos\left(\omega t' - \omega t + \frac{\omega r}{c}\right) \tag{4.107}$$

Next, forming the $(l + 1)$ derivative of the magnetic form factor under the integrals in terms of the magnetic radii leads to:

$$M_{lm}^{(l+1)}\left(-\frac{\omega^2}{c^2}, t'\right) = \sum_{n=0}^{\infty} \frac{\left(-\frac{\omega^2}{c^2}\right)^n}{n!} M_{lm}^{(n)(l+1)}(0, t') \tag{4.108}$$

By introducing double-superscript quantities $M_{lm}^{(n)(l+1)}(0, t')$ in (4.14), (4.108) can be described in terms of Taylor series:

$$\int_0^\infty d\omega\omega^{l+1} M_{lm}\left(-\frac{\omega^2}{c^2},\omega\right)\sin\left(\omega\left(t-\frac{r}{c}\right)\right)$$
$$=\frac{(-1)^{l/2+1}}{2}\frac{\pi}{2}\sum_{n=0}^{\infty}\frac{1}{n!c^{2n}}M_{lm}^{(n)(2n+l+1)}\left(0,t-\frac{r}{c}\right) \qquad (4.109)$$

In a similar fashion, one can extract the followings for the electric and toroidal form factors:

$$\int_0^\infty d\omega\omega^{l+1} Q_{lm}(0,\omega)\cos\left(\omega\left(t-\frac{r}{c}\right)\right)$$
$$=\frac{\pi(-1)^{l/2}}{2}Q_{lm}^{(0)(l+1)}(0,\omega)\left(0,t-\frac{r}{c}\right) \qquad (4.110)$$

$$\int_0^\infty d\omega\omega^{l+2} T_{lm}\left(-\frac{\omega^2}{c^2},\omega\right)\cos\left(\omega\left(t-\frac{r}{c}\right)\right)$$
$$=\frac{(-1)^{l/2+1}}{2}\frac{\pi}{2}\sum_{n=0}^{\infty}\frac{1}{n!c^{2n}}T_{lm}^{(n)(2n+l+2)}\left(0,t-\frac{r}{c}\right) \qquad (4.111)$$

These set of equations complete the calculation of the $l = even$ part of $\vec{B}(\vec{r},t)$ in $O(1/r)$.

On the other hand, in the $l = odd$ regime ($l = 2\,k + 1$), the analysis will be similar to the *even* limit, with minor modifications, hence, the final results remain unchanged for $\vec{B}(\vec{r},t)$ at $O(1/r)$. In the wave zone, one can define the electric field using:

$$\vec{E}(\vec{r},t) = -\frac{\vec{r}}{r}\times\vec{B}(\vec{r},t) \qquad (4.112)$$

Consequently, it would be possible to obtain the following expressions for the electric and magnetic fields at large distances in the order of $O(1/r)$, described in terms of the double derivatives of the form factors. Here, the first superscript shows the order of the derivation with respect to the first argument of the form factor at zero value, and the second superscript indicates the order of derivation with respect to the second argument. Hence:

$$\vec{E}(\vec{r},t) \overset{O(1/r)}{\sim} \frac{1}{r}\sum_{l,m}\frac{1}{c^{l+1}}\frac{\sqrt{4\pi(2l+1)(l+1)}}{\sqrt{l}(2l+1)!!}\left\{-Q_{lm}^{(0)(l+1)}\left(0,t-\frac{r}{c}\right)\right.$$
$$\times\left[\frac{\sqrt{l}}{\sqrt{2l+1}}\vec{Y}_{ll+1m}(\vec{n})+\frac{\sqrt{l+1}}{\sqrt{2l+1}}\vec{Y}_{ll-1m}(\vec{n})\right]$$

$$
\begin{aligned}
&- i\sum_{n=0}^{\infty}\frac{1}{n!c^{2n}}M_{lm}^{(n)(2n+l+1)}\left(0,t-\frac{r}{c}\right)\vec{Y}_{llm}(\vec{n})\\
&+\frac{1}{c}\sum_{n=0}^{\infty}\frac{1}{n!c^{2n}}T_{lm}^{(n)(2n+l+2)}\left(0,t-\frac{r}{c}\right)\vec{Y}_{llm}(\vec{n})\\
&\times\left[\frac{\sqrt{l}}{\sqrt{2l+1}}\vec{Y}_{ll+1m}(\vec{n})+\frac{\sqrt{l+1}}{\sqrt{2l+1}}\vec{Y}_{ll-1m}(\vec{n})\right]\Bigg\}
\end{aligned}
\tag{4.113}
$$

$$
\begin{aligned}
\vec{B}(\vec{r},t)\overset{O(1/r)}{\sim}\frac{1}{r}\sum_{l,m}\frac{1}{c^{l+1}}\frac{\sqrt{4\pi(2l+1)(l+1)}}{\sqrt{l}(2l+1)!!}\Bigg\{&-Q_{lm}^{(0)(l+1)}\left(0,t-\frac{r}{c}\right)\vec{Y}_{llm}(\vec{n})\\
&+i\sum_{n=0}^{\infty}\frac{1}{n!c^{2n}}M_{lm}^{(n)(2n+l+1)}\left(0,t-\frac{r}{c}\right)\\
&\times\left[\frac{\sqrt{l}}{\sqrt{2l+1}}\vec{Y}_{ll+1m}(\vec{n})+\frac{\sqrt{l+1}}{\sqrt{2l+1}}\vec{Y}_{ll-1m}(\vec{n})\right]\\
&+\frac{i}{c}\sum_{n=0}^{\infty}\frac{1}{n!c^{2n}}T_{lm}^{(n)(2n+l+2)}\left(0,t-\frac{r}{c}\right)\vec{Y}_{llm}(\vec{n})\Bigg\}
\end{aligned}
\tag{4.114}
$$

In the upcoming sections, for the calculation of the angular momentum loss, we need to define the following terms: $\vec{n}\vec{E}$, $\vec{n}\vec{B}$, $\vec{n}\times\vec{E}$, $\vec{n}\times\vec{B}$ at large distances. Since $\vec{n}\vec{E}$ and $\vec{n}\vec{B}$ are zero at the order of $O(1/r)$, these components should be evaluated at the other of $O(1/r^2)$. This enables us to obtain the first relevant nonvanishing contributions.

To begin with, we limit our studies to the first order $O(1/r)$. Therefore:

$$
\vec{n}\vec{F}_{lm(\omega/c)}^{(n)}(\vec{r})=\vec{n}\vec{N}_{lm(\omega/c)}^{(n)}(\vec{r})=0 \tag{4.115}
$$

Although $(\vec{Y}_{llm})_r=0$, using the (−) and (+) superscripts, one has:

$$
\begin{aligned}
\left[Y_{ll+1m}(\vec{n})\right]_r&=-\frac{\sqrt{l+1}}{\sqrt{2l+1}}Y_{lm}(\vec{n})\\
\left[Y_{ll-1m}(\vec{n})\right]_r&=\frac{\sqrt{l}}{\sqrt{2l+1}}Y_{lm}(\vec{n})
\end{aligned}
\tag{4.116}
$$

one can write:

$$
\vec{n}\vec{F}_{lm(\omega/c)}^{(+)}(\vec{r})=\frac{\sqrt{l(l+1)}}{(2l+1)}\left[-j_{l+1}\left(\frac{\omega}{c}r\right)+j_{l-1}\left(\frac{\omega}{c}r\right)\right]Y_{lm}(\vec{n}) \tag{4.117}
$$

$$
\vec{n}\vec{F}_{lm(\omega/c)}^{(-)}(\vec{r})=\frac{1}{(2l+1)}\left[lj_{l-1}\left(\frac{\omega}{c}r\right)+(l+1)j_{l+1}\left(\frac{\omega}{c}r\right)\right]Y_{lm}(\vec{n}) \tag{4.118}
$$

$$\vec{n}\vec{N}^{(+)}_{lm(\omega/c)}(\vec{r}) = \frac{\sqrt{l(l+1)}}{(2l+1)}\left[-n_{l+1}\left(\frac{\omega}{c}r\right) + n_{l-1}\left(\frac{\omega}{c}r\right)\right]Y_{lm}(\vec{n}) \tag{4.119}$$

$$\vec{n}\vec{N}^{(-)}_{lm(\omega/c)}(\vec{r}) = \frac{1}{(2l+1)}\left[ln_{l-1}\left(\frac{\omega}{c}r\right) + (l+1)n_{l+1}\left(\frac{\omega}{c}r\right)\right]Y_{lm}(\vec{n}) \tag{4.120}$$

Now, by utilizing the asymptotical behavior of j_l and n to the order of $O(1/r)$:

$$\vec{n}\vec{F}^{(+)}_{lm(\omega/c)}(\vec{r}) \underset{r\to\infty}{\overset{O(1/r)}{\sim}} 0 \tag{4.121}$$

$$\vec{n}\vec{F}^{(-)}_{lm(\omega/c)}(\vec{r}) \underset{r\to\infty}{\overset{O(1/r)}{\sim}} -4\pi i^{l+1}\frac{\cos\left(\frac{\omega}{c}r - l\frac{\pi}{2}\right)}{\frac{\omega}{c}r}Y_{lm}(\vec{n}) \tag{4.122}$$

$$\vec{n}\vec{N}^{(+)}_{lm(\omega/c)}(\vec{r}) \underset{r\to\infty}{\overset{O(1/r)}{\sim}} 0 \tag{4.123}$$

$$\vec{n}\vec{N}^{(-)}_{lm(\omega/c)}(\vec{r}) \underset{r\to\infty}{\overset{O(1/r)}{\sim}} -4\pi i^{l+1}\frac{\sin\left(\frac{\omega}{c}r - l\frac{\pi}{2}\right)}{\frac{\omega}{c}r}Y_{lm}(\vec{n}) \tag{4.124}$$

With these assumptions, (2.88) and (2.89) at $O(1/r)$ can be approximated as:

$$\frac{\vec{r}}{r}\vec{E}(\vec{r},t) \underset{r\to\infty}{\overset{O(1/r)}{\sim}} 0 \tag{4.125}$$

$$\frac{\vec{r}}{r}\vec{B}(\vec{r},t) \underset{r\to\infty}{\overset{O(1/r)}{\sim}} 0 \tag{4.126}$$

By considering the expressions for $\vec{E}(\vec{r},t)$ and $\vec{B}(\vec{r},t)$ to the order of $O(1/r)$, for determining $\vec{n}\times\vec{E}$ and $\vec{n}\times\vec{B}$ to the same order, one requires cross products between $\vec{n} = \vec{r}/r$ and the appearing vector spherical harmonics. Using the following unit vectors:$\vec{e}_r = \vec{n} = \vec{r}/r$, $\vec{e}_\theta$, $\vec{e}_\varphi$, one can obtain:

$$\begin{aligned}\vec{e}_r \times \vec{Y}_{llm}(\vec{n}) &= \vec{e}_\varphi(\vec{Y}_{llm})_\theta - \vec{e}_\theta \times (\vec{Y}_{llm})_\varphi \\ &= i\frac{\sqrt{2l+1}}{\sqrt{l+1}}\left[\vec{Y}_{ll-1m} - \vec{e}_r\frac{\sqrt{l}}{\sqrt{2l+1}}\vec{Y}_{lm}\right] \\ &= i\frac{\sqrt{2l+1}}{\sqrt{l}}\left[\vec{Y}_{ll+1m} + \vec{e}_r\frac{\sqrt{l}}{\sqrt{2l+1}}\vec{Y}_{lm}\right]\end{aligned} \tag{4.127}$$

since

$$\left(\vec{Y}_{llm}\right)_\theta = i\frac{\sqrt{2l+1}}{\sqrt{l+1}}\left(\vec{Y}_{ll-1m}\right)_\varphi = i\frac{\sqrt{2l+1}}{\sqrt{l}}\left(\vec{Y}_{ll+1m}\right)_\varphi$$

$$\left(\vec{Y}_{llm}\right)_{\varphi} = -i\frac{\sqrt{2l+1}}{\sqrt{l+1}}\left(\vec{Y}_{ll-1m}\right)_{\theta} = -i\frac{\sqrt{2l+1}}{\sqrt{l}}\left(\vec{Y}_{ll+1m}\right)_{\theta} \tag{4.128}$$

The terms with $\vec{e}_r\vec{Y}_{llm}$ in (4.127) can be simply removed, therefore:

$$\vec{n}\times\vec{Y}_{llm}(\vec{n}) = i\frac{\sqrt{l+1}}{\sqrt{2l+1}}\vec{Y}_{ll-1m}(\vec{n}) + i\frac{\sqrt{l}}{\sqrt{2l+1}}\vec{Y}_{ll+1m}(\vec{n}) \tag{4.129}$$

Moreover,

$$\vec{e}_r\times\vec{Y}_{ll+1m}(\vec{n}) = \vec{e}_{\varphi}\left(\vec{Y}_{ll+1m}\right)_{\theta} - \vec{e}_{\theta}\left(\vec{Y}_{ll+1m}\right)_{\varphi} = i\frac{\sqrt{l}}{\sqrt{2l+1}}\vec{Y}_{llm}(\vec{n}) \tag{4.130}$$

because

$$\begin{aligned}\left(\vec{Y}_{ll+1m}\right)_{\theta} &= i\frac{\sqrt{l}}{\sqrt{2l+1}}\left(\vec{Y}_{llm}\right)_{\varphi}\\ \left(\vec{Y}_{ll+1m}\right)_{\varphi} &= -i\frac{\sqrt{l}}{\sqrt{2l+1}}\left(\vec{Y}_{llm}\right)_{\theta}\end{aligned} \tag{4.131}$$

and similarly:

$$\vec{e}_r\times\vec{Y}_{ll-1m}(\vec{n}) = i\frac{\sqrt{l+1}}{\sqrt{2l+1}}\vec{Y}_{llm}(\vec{n}) \tag{4.132}$$

Using the cross products of (4.130)–(4.132), the following universal relations can be obtained for $\vec{n}\times\vec{F}^{(\lambda)}_{lm(\omega/c)}(\vec{r})$ and $\vec{n}\times\vec{N}^{(\lambda)}_{lm(\omega/c)}(\vec{r})$ ($\vec{e}_r = \vec{n} = \vec{r}/r$, $\lambda = 0, \pm$):

$$\vec{n}\times\vec{F}^{(0)}_{lm(\omega/c)}(\vec{r}) = j_l\left(\frac{\omega}{c}r\right)\left[i\frac{\sqrt{l+1}}{\sqrt{2l+1}}\vec{Y}_{ll-1m}(\vec{n}) + i\frac{\sqrt{l}}{\sqrt{2l+1}}\vec{Y}_{ll+1m}(\vec{n})\right] \tag{4.133}$$

$$\vec{n}\times\vec{F}^{(+)}_{lm(\omega/c)}(\vec{r}) = \frac{i}{2l+1}\left[l j_{l+1}\left(\frac{\omega}{c}r\right) + (l+1)j_{l-1}\left(\frac{\omega}{c}r\right)\right]\vec{Y}_{llm}(\vec{n}) \tag{4.134}$$

$$\vec{n}\times\vec{F}^{(-)}_{lm(\omega/c)}(\vec{r}) = \frac{i\sqrt{l(l+1)}}{2l+1}\left[j_{l-1}\left(\frac{\omega}{c}r\right) - j_{l+1}\left(\frac{\omega}{c}r\right)\right]\vec{Y}_{llm}(\vec{n}) \tag{4.135}$$

$$\vec{n}\times\vec{N}^{(0)}_{lm(\omega/c)}(\vec{r}) = n_l\left(\frac{\omega}{c}r\right)\left[i\frac{\sqrt{l+1}}{\sqrt{2l+1}}\vec{Y}_{ll-1m}(\vec{n}) + i\frac{\sqrt{l}}{\sqrt{2l+1}}\vec{Y}_{ll+1m}(\vec{n})\right] \tag{4.136}$$

$$\vec{n}\times\vec{N}^{(+)}_{lm(\omega/c)}(\vec{r}) = \frac{i}{2l+1}\left[l n_{l+1}\left(\frac{\omega}{c}r\right) + (l+1)n_{l-1}\left(\frac{\omega}{c}r\right)\right]\vec{Y}_{llm}(\vec{n}) \tag{4.137}$$

$$\vec{n} \times \vec{N}^{(-)}_{lm(\omega/c)}(\vec{r}) = \frac{i\sqrt{l(l+1)}}{2l+1}\left[n_{l-1}\left(\frac{\omega}{c}r\right) - n_{l+1}\left(\frac{\omega}{c}r\right)\right]\vec{Y}_{llm}(\vec{n}) \tag{4.138}$$

Next, using the large r feature of the spherical Bessel functions to the order $1/r$, the preferred asymptotical behavior of (4.133)–(4.138) to the order $O(1/r)$ of the cross products $\vec{n} \times \vec{F}^{(\lambda)}_{lm(\omega/c)}(\vec{r})$ and $\vec{n} \times \vec{N}^{(\lambda)}_{lm(\omega/c)}(\vec{r})$ $(\vec{n} = \vec{r}/r)$ will be:

$$\vec{n} \times \vec{F}^{(0)}_{lm(\omega/c)}(\vec{r}) \overset{O(1/r)}{\sim} \frac{4\pi i^{l+1}}{\sqrt{2l+1}}\left[\sqrt{l+1}\vec{Y}_{ll-1m}(\vec{n}) + \sqrt{l}\vec{Y}_{ll+1m}(\vec{n})\right]\frac{\sin\left(\frac{\omega}{c}r - l\frac{\pi}{2}\right)}{\frac{\omega}{c}r} \tag{4.139}$$

$$\vec{n} \times \vec{F}^{(+)}_{lm(\omega/c)}(\vec{r}) \overset{O(1/r)}{\sim} 4\pi i^l \vec{Y}_{llm}(\vec{n})\frac{\cos\left(\frac{\omega}{c}r - l\frac{\pi}{2}\right)}{\frac{\omega}{c}r} \tag{4.140}$$

$$\vec{n} \times \vec{F}^{(-)}_{lm(\omega/c)}(\vec{r}) \overset{O(1/r)}{\sim} 0 \tag{4.141}$$

$$\vec{n} \times \vec{N}^{(0)}_{lm(\omega/c)}(\vec{r}) \overset{O(1/r)}{\sim} -\frac{4\pi i^{l+1}}{\sqrt{2l+1}}\left[\sqrt{l+1}\vec{Y}_{ll-1m}(\vec{n}) + \sqrt{l}\vec{Y}_{ll+1m}(\vec{n})\right]\frac{\cos\left(\frac{\omega}{c}r - l\frac{\pi}{2}\right)}{\frac{\omega}{c}r} \tag{4.142}$$

$$\vec{n} \times \vec{N}^{(+)}_{lm(\omega/c)}(\vec{r}) \overset{O(1/r)}{\sim} 4\pi i^l \vec{Y}_{llm}(\vec{n})\frac{\sin\left(\frac{\omega}{c}r - l\frac{\pi}{2}\right)}{\frac{\omega}{c}r} \tag{4.143}$$

$$\vec{n} \times \vec{N}^{(-)}_{lm(\omega/c)}(\vec{r}) \overset{O(1/r)}{\sim} 0 \tag{4.144}$$

Taking advantage of (4.139)–(4.144), the targeted asymptotical behavior of $(\vec{r}/r) \times \vec{E}(\vec{r}, t)$ at large distances can be described in terms of the electric, magnetic, and toroidal form factors:

$$\begin{aligned}
\left(\frac{\vec{r}}{r}\right) \times \vec{E}(\vec{r}, t) \overset{O(1/r)}{\sim} & -\frac{4}{\sqrt{\pi}}\frac{1}{r}\int_0^\infty d\omega \sum_{l,m} \frac{\omega^{l+1}}{c^{l+1}}\frac{\sqrt{l+1}}{\sqrt{l}(2l+1)!!} \\
& \times \left\{\sqrt{l+1}M_{lm}\left(-\frac{\omega^2}{c^2}, \omega\right)\right. \\
& \times \sin\left(\omega t - \frac{\omega}{c}r + l\frac{\pi}{2}\right)\vec{Y}_{ll-1m}(\vec{n}) + \sqrt{l}M_{lm}\left(-\frac{\omega^2}{c^2}, \omega\right) \\
& \times \sin\left(\omega t - \frac{\omega}{c}r + l\frac{\pi}{2}\right)\vec{Y}_{ll+1m}(\vec{n}) + i\sqrt{l+1}\left[Q_{lm}(0, \omega)\right. \\
& \left.\left.+\frac{\omega}{c}T_{lm}\left(-\frac{\omega^2}{c^2}, \omega\right)\right]\cos\left(\omega t - \frac{\omega}{c}r + l\frac{\pi}{2}\right)\vec{Y}_{llm}(\vec{n})\right\}
\end{aligned} \tag{4.145}$$

Analogously, the behavior of magnetic field at large distances can be attained as following:

$$
\begin{aligned}
(\frac{\vec{r}}{r}) \times \vec{B}(\vec{r},t) \overset{O(1/r)}{\sim} & -\frac{4}{\sqrt{\pi}}\frac{1}{r}\int\limits_0^\infty d\omega \sum_{l,m} \frac{\omega^{l+1}}{c^{l+1}} \frac{\sqrt{l+1}}{\sqrt{l}(2l+1)!!} \\
& \times \left\{ i\sqrt{2l+1} M_{lm}\left(-\frac{\omega^2}{c^2},\omega\right) \right. \\
& \times \sin\left(\omega t - \frac{\omega}{c}r + l\frac{\pi}{2}\right)\vec{Y}_{llm}(\vec{n}) \\
& - \left[Q_{lm}(0,\omega) + \frac{\omega}{c}T_{lm}\left(-\frac{\omega^2}{c^2},\omega\right)\right] \\
& \left. \times \cos\left(\omega t - \frac{\omega}{c}r + l\frac{\pi}{2}\right)\left[\sqrt{l+1}\vec{Y}_{ll-1m}(\vec{n}) + \sqrt{l}\vec{Y}_{ll+1m}(\vec{n})\right]\right\}
\end{aligned} \tag{4.146}
$$

The equations above, (1.145) and (1.146), can be employed to find the expressions for $\vec{n} \times \vec{E}$ and $\vec{n} \times \vec{B}$ to the order $O(1/r)$, in terms of the double-superscript quantities $Q_{lm}^{(n)(\upsilon)}(0,t)$, $M_{lm}^{(n)(\upsilon)}(0,t)$, $T_{lm}^{(n)(\upsilon)}(0,t)$, as:

$$
\begin{aligned}
(\frac{\vec{r}}{r}) \times \vec{E}(\vec{r},t) \overset{O(1/r)}{\sim} & \frac{2i\sqrt{\pi}}{r}\sum_{l,m}\frac{1}{c^{l+1}}\frac{\sqrt{(2l+1)(l+1)}}{\sqrt{l}(2l+1)!!}\left\{Q_{lm}^{(0)(l+1)}\left(0,t-\frac{r}{c}\right)\right. \\
& \left. + \sum_{n=0}^{\infty}\frac{1}{n!c^{2n+1}}T_{lm}^{(n)(l+2n+2)}\left(0,t-\frac{r}{c}\right)\right\}\vec{Y}_{llm}(\vec{n}) + \frac{2\sqrt{\pi}}{r} \\
& \times \sum_{lm}\sum_{n=0}^{\infty}\frac{1}{n!c^{l+2n+1}}\frac{\sqrt{(l+1)}}{\sqrt{l}(2l+1)!!}M_{lm}^{(n)(l+2n+1)}\left(0,t-\frac{r}{c}\right) \\
& + \left[\sqrt{(l+1)}\vec{Y}_{ll-1m}(\vec{n}) + \sqrt{l}\vec{Y}_{ll+1m}(\vec{n})\right]
\end{aligned} \tag{4.147}
$$

Analogously, the magnetic field can be found in the same way as above:

$$
\begin{aligned}
(\frac{\vec{r}}{r}) \times \vec{B}(\vec{r},t) \overset{O(1/r)}{\sim} & \frac{2\sqrt{\pi}}{r}\frac{1}{r}\int\limits_0^\infty d\omega\sum_{l,m}\frac{\sqrt{l+1}}{\sqrt{l}(2l+1)!!}\left\{\frac{1}{c}Q_{lm}^{(0)(l+1)}\left(0,t-\frac{r}{c}\right)\right. \\
& \left. - \sum_{n=0}^{\infty}\frac{1}{n!c^{l+2n+2}}T_{lm}^{(n)(l+2n+2)}\left(0,t-\frac{r}{c}\right)\right\} \\
& \times \left[\sqrt{l+1}\vec{Y}_{ll-1m}(\vec{n}) + \sqrt{l}\vec{Y}_{ll+1m}(\vec{n})\right] + \frac{2i\sqrt{\pi}}{r} \\
& \times \sum_{lm}\sum_{n=0}^{\infty}\frac{\sqrt{(2l+1)(l+1)}}{\sqrt{l}(2l+1)!!}\frac{1}{n!c^{l+2n+1}} \\
& \times M_{lm}^{(n)(l+2n+1)}\left(0,t-\frac{r}{c}\right)\vec{Y}_{llm}(\vec{n})
\end{aligned} \tag{4.148}
$$

We further consider the $1/r^2$ order to evaluate $\vec{n} \cdot \vec{E}$ and $\vec{n} \cdot \vec{B}$. To this end, the first nonvanishing contributions to these quantities should be extracted. Thereby, the terms in the asymptotical behavior of the spherical Bessel functions as compared with (4.86)–(4.91) can be formulated as:

$$j_l(x) \underset{x\to\infty}{\overset{O(1/x^2)}{\sim}} 4\pi i^l \frac{\sin\left(x - l\frac{\pi}{2}\right)}{x} + 2\pi i^l l(l+1) \frac{\cos\left(x - l\frac{\pi}{2}\right)}{x^2} \tag{4.149}$$

$$n_l(x) \underset{x\to\infty}{\overset{O(1/x^2)}{\sim}} -4\pi i^l \frac{\cos\left(x - l\frac{\pi}{2}\right)}{x} + 2\pi i^l l(l+1) \frac{\sin\left(x - l\frac{\pi}{2}\right)}{x^2} \tag{4.150}$$

which obey the following:

$$\begin{aligned} j_{l+1/2}(x) = \sqrt{\frac{2}{\pi x}} \Bigg\{ &\sin\left(x - l\frac{\pi}{2}\right) + \sum_{k=0}^{l/2} \frac{(-1)^k (l+2k)!}{(2k)!(l-2k)!(2x)^{2k}} \\ &+ \cos\left(x - l\frac{\pi}{2}\right) \sum_{k=0}^{(l-1)/2} \frac{(-1)^k (l+2k+1)!}{(2k+1)!(l+2k-1)!(2x)^{2k+1}} \Bigg\} \end{aligned} \tag{4.151}$$

and relations:

$$\begin{cases} j_l(x) = (2\pi)^{3/2} \dfrac{i^l}{\sqrt{x}} J_{l+1/2}(x) \\ n_l(x) = (2\pi)^{3/2} \dfrac{i^l}{\sqrt{x}} N_{l+1/2}(x) \\ N_{l+1/2}(x) = (-1)^{l-1} J_{-(l+1/2)}(x) \end{cases} \tag{4.152}$$

Considering (4.149) and (4.150) at the order of $O(1/r^2)$ (where $\vec{n} = \vec{r}/r$):

$$\vec{n}.\vec{F}^{(+)}_{lm(\omega/c)}(\vec{r}) \sim -4\pi i^{l+1} \sqrt{l(l+1)} \frac{\sin\left(\frac{\omega}{c} r - l\frac{\pi}{2}\right)}{\left(\frac{\omega}{c} r\right)^2} Y_{lm}(\vec{r}) \tag{4.153}$$

$$\begin{aligned} \vec{n}.\vec{F}^{(-)}_{lm(\omega/c)}(\vec{r}) \sim \Bigg\{ &-4\pi i^{l+1} \frac{\cos\left(\frac{\omega}{c} r - l\frac{\pi}{2}\right)}{\left(\frac{\omega}{c} r\right)} \\ &+ 2\pi i^{l+1} (l^2 + l + 2) \frac{\sin\left(\frac{\omega}{c} r - l\frac{\pi}{2}\right)}{\left(\frac{\omega}{c} r\right)^2} \Bigg\} Y_{lm}(\vec{r}) \end{aligned} \tag{4.154}$$

$$\vec{n}.\vec{N}^{(+)}_{lm(\omega/c)}(\vec{r}) \sim 4\pi i^{l+1} \sqrt{l(l+1)} \frac{\cos\left(\frac{\omega}{c} r - l\frac{\pi}{2}\right)}{\left(\frac{\omega}{c} r\right)^2} Y_{lm}(\vec{r}) \tag{4.155}$$

$$\vec{n}.\vec{N}^{(-)}_{lm(\omega/c)}(\vec{r}) \sim \Bigg\{ -4\pi i^{l+1} \sin \frac{\cos\left(\frac{\omega}{c} r - l\frac{\pi}{2}\right)}{\left(\frac{\omega}{c} r\right)}$$

$$-2\pi i^{l+1}(l^2+l+2)\frac{\sin\left(\frac{\omega}{c}r-l\frac{\pi}{2}\right)}{\left(\frac{\omega}{c}r\right)^2}\Bigg\}Y_{lm}(\vec{r}) \tag{4.156}$$

Therefore, from (2.88), (2.89), and (4.156), the leading contributions to $\vec{n}\cdot\vec{E}$ and $\vec{n}\cdot\vec{B}$ on the order of $O(1/r^2)$ can be found as:

$$\begin{aligned}(\frac{\vec{r}}{r})\cdot\vec{E}(\vec{r},t) \overset{O(1/r^2)}{\sim} & \frac{4}{r^2\sqrt{\pi}}\int_0^\infty d\omega\sum_{l,m}\frac{1}{c^{l+1}}\frac{(l+1)\sqrt{(2l+1)}}{(2l+1)!!}\left(\frac{\omega}{c}\right)^l \\ & \times\left\{Q_{lm}(0,\omega)+\frac{\omega}{c}T_{lm}\left(-\frac{\omega^2}{c^2},\omega\right)\right\} \\ & \times\sin\left(\omega t-\frac{\omega}{c}r+l\frac{\pi}{2}\right)\vec{Y}_{lm}(\vec{n})\end{aligned} \tag{4.157}$$

and

$$\begin{aligned}(\frac{\vec{r}}{r})\cdot\vec{B}(\vec{r},t) \overset{O(1/r^2)}{\sim} & -\frac{4}{r^2\sqrt{\pi}}\int_0^\infty d\omega\sum_{l,m}\frac{(l+1)\sqrt{(2l+1)}}{(2l+1)!!}\left(\frac{\omega}{c}\right)^l \\ & \times\left\{M_{lm}\left(-\frac{\omega^2}{c^2},\omega\right)\cos\left(\omega t-\frac{\omega}{c}r+l\frac{\pi}{2}\right)\right\}\times\vec{Y}_{lm}(\vec{n})\end{aligned} \tag{4.158}$$

The descriptions of $\vec{n}\cdot\vec{E}$ and $\vec{n}\cdot\vec{B}$ to the order of $O(1/r^2)$ in terms of time derivatives of radii (i.e., double-superscript quantities) can be obtained through the use of (4.109)–(4.111) and considering the *even* and *odd* contributions to the sum over l. Hence:

$$\begin{aligned}(\frac{\vec{r}}{r})\cdot\vec{E}(\vec{r},t) \overset{O(1/r^2)}{\sim} & \frac{2\sqrt{\pi}}{r^2}\sum_{l,m}\frac{1}{c^l}\frac{(l+1)\sqrt{2l+1}}{(2l+1)!!}\left\{Q_{lm}^{(0)(l)}\left(0,t-\frac{r}{c}\right)\right. \\ & \left.-\sum_{n=0}^\infty\frac{1}{n!c^{2n+1}}T_{lm}^{(n)(l+2n+1)}\left(0,t-\frac{r}{c}\right)\right\}\vec{Y}_{lm}(\vec{n})\end{aligned} \tag{4.159}$$

and

$$\begin{aligned}(\frac{\vec{r}}{r})\cdot\vec{B}(\vec{r},t) \overset{O(1/r^2)}{\sim} & -\frac{2\sqrt{\pi}}{r^2}\sum_{l,m}\frac{1}{c^l}\frac{(l+1)\sqrt{2l+1}}{(2l+1)!!} \\ & \times\left\{\sum_{n=0}^\infty\frac{1}{n!c^{2n}}M_{lm}^{(n)(l+2n)}\left(0,t-\frac{r}{c}\right)\right\}\vec{Y}_{lm}(\vec{n})\end{aligned} \tag{4.160}$$

In continue, establishing the required quantities for the calculation of the angular momentum loss, recoil force, and radiation intensity, we further explain the definition of these features in terms of the multipole content of the most general type of source.

4.2 Radiation Intensity

To define the radiation intensity for a general distribution of charges and currents, in terms of multipole form factors, one should consider $M_{lm}^{(n)(l+q)}(0, t - r/c)$, in which M_{lm} stands for any multipole form factor (e.g.,$Q_{lm}(k^2, t)$, $M_{lm}(k^2, t)$, $T_{lm}(k^2, t)$), (n) indicates the derivative with respect to $(-k^2)$ at $k = 0$, and $(l + q)$ denotes the derivative with respect to t at $t - r/c$. Therefore, the Poynting vector can be written as:

$$\vec{S} = \frac{c}{4\pi}(\vec{E} \times \vec{B}) \tag{4.161}$$

With this way, one can express the radiation intensity as the surface integral of the radial component of $\vec{S}(S_r)$ over a sphere of charge radius r (Landau and Lifshitz 1993):

$$I = r^2 \int d\Omega\, S_r = \frac{cr^2}{4\pi} \int d\Omega(\vec{E} \times \vec{B})_r \tag{4.162}$$

Now, we can evaluate the electric and magnetic fields to the order $O(1/r)$ using (4.113) and (4.114). For this purpose, one needs to assess the surface integral over the r component of the vector:

$$\begin{aligned}
\vec{E} \times \vec{B} = \frac{4\pi}{r^2} \sum_{l,m,l',m'} \frac{1}{c^{l+l'+2}} \frac{\sqrt{(2l+1)(l+1)(2l'+1)(l'+1)}}{\sqrt{l}\sqrt{l'}(2l+1)!!(2l'+1)!!} \\
\times \left\{ -i \sum_{n=0}^{\infty} \frac{1}{n!c^{2n}} M_{lm}^{(n)(l+2n+1)} \vec{Y}_{llm} \right. \\
- Q_{lm}^{(0)(l+1)} \left[\frac{\sqrt{l}}{\sqrt{2l+1}} \vec{Y}_{ll+1m} + \frac{\sqrt{l+1}}{\sqrt{2l+1}} \vec{Y}_{ll-1m} \right] \\
+ \frac{1}{c} \sum_{n=0}^{\infty} \frac{1}{n!c^{2n}} T_{lm}^{(n)(l+2n+2)} \\
\times \left. \left[\frac{\sqrt{l}}{\sqrt{2l+1}} \vec{Y}_{ll+1m} + \frac{\sqrt{l+1}}{\sqrt{2l+1}} \vec{Y}_{ll-1m} \right] \right\} \\
\times \left\{ \sum_{n'=0}^{\infty} \frac{1}{n'!c^{2n'}} M_{l'm'}^{(n')(l'+2n'+1)} \left[\frac{\sqrt{l'}}{\sqrt{2l'+1}} \vec{Y}_{l'l'+1m'} \right. \right.
\end{aligned}$$

$$+\frac{\sqrt{l'+1}}{\sqrt{2l'+1}}\vec{Y}_{l'l'-1m'}\Bigg] - i\,Q^{(0)(l'+1)}_{l'm'}\vec{Y}_{l'l'm'}$$
$$+\frac{i}{c}\sum_{n'=0}^{\infty}\frac{1}{n'!c^{2n'}}T^{(n')(l'+2n'+2)}_{l'm'}\vec{Y}_{l'l'm'}\Bigg\} \tag{4.163}$$

In the equation above, the argument of the vector spherical harmonics $\vec{Y}$ is $\vec{n} = \vec{r}/r$. The surface integration in (4.162) across the cross products of spherical vector harmonics is accomplished through simplifying them to the dot products of $\vec{Y}$. Therefore, using the following equations:

$$\left(\vec{Y}_{l'l'm'}\times\vec{Y}_{ll+1m}\right)_r = -i\frac{\sqrt{l}}{\sqrt{2l+1}}\vec{Y}_{l'l'm'}\cdot\vec{Y}_{llm} \tag{4.164}$$

$$\left(\vec{Y}_{l'l'+1m'}\times\vec{Y}_{ll-1m}\right)_r = i\frac{\sqrt{l'}}{\sqrt{2l'+1}}\vec{Y}_{l'l'm'}\cdot\vec{Y}_{ll-1m} \tag{4.165}$$

and

$$\int d\Omega\left(\vec{Y}_{ll+1m}\times\vec{Y}_{l'l'm'}\right)_r = -i(-1)^m\frac{\sqrt{l}}{\sqrt{2l+1}}\delta_{ll'}\delta_{m,m'} \tag{4.166}$$

$$\int d\Omega\left(\vec{Y}_{ll-1m}\times\vec{Y}_{l'l'm'}\right)_r = -i(-1)^m\frac{\sqrt{l+1}}{\sqrt{2l+1}}\delta_{ll'}\delta_{m,m'} \tag{4.167}$$

one can now obtain the intensity of radiation in terms of the derivatives of electric, magnetic, and toroidal components with respect to $-k^2$ and t:

$$I_{rad} = \sum_{l,m}\frac{1}{c^{2l+1}}\frac{(l+1)}{l(2l-1)!!(2l+1)!!}\Bigg\{\left|Q^{(0)(l+1)}_{lm}\left(0,t-\frac{r}{c}\right)\right|^2$$
$$-\sum_{n}\frac{1}{n!c^{2n+1}}\left[Q^{(0)(l+1)}_{lm}\left(0,t-\frac{r}{c}\right)T^{(n)(l+2n+2)}_{lm}*\left(0,t-\frac{r}{c}\right)\right.$$
$$\left.+Q^{(0)(l+1)}_{lm}*\left(0,t-\frac{r}{c}\right)T^{(n)(l+2n+2)}_{lm}\left(0,t-\frac{r}{c}\right)\right]$$
$$+\sum_{n,n'}\frac{1}{n!n'!c^{2n+2n'}}\left[M^{(n)(l+2n+1)}_{lm}\left(0,t-\frac{r}{c}\right)M^{(n')(l+2n'+1)}_{lm}*\left(0,t-\frac{r}{c}\right)\right.$$
$$\left.+\frac{1}{c^2}T^{(n)(l+2n+2)}_{lm}\left(0,t-\frac{r}{c}\right)T^{(n')(l+2n'+2)}_{lm}*\left(0,t-\frac{r}{c}\right)\right]\Bigg\} \tag{4.168}$$

Here, it is noteworthy to mention that (4.168) explains the radiation intensity for a general type of source based on $Q_{lm}(k^2,t)$, $M_{lm}(k^2,t)$, and $T_{lm}(k^2,t)$ form factors for an arbitrary time-dependence. The first superscript implies the order of derivation with respect to the first argument, $-k^2$, and the second superscript indicates the order of derivation with respect to the second argument, t. Importantly, all terms in (4.168) are real and equal to their complex conjugates. Thus, the radiation intensity I_{rad} can be written in much simpler and compact format:

$$I_{rad} = \sum_{l,m} \frac{1}{c^{2l+1}} \frac{(l+1)}{l(2l-1)!!(2l+1)!!} \Bigg\{ \Bigg| Q_{lm}^{(0)(l+1)}\left(0, t - \frac{r}{c}\right) - \frac{1}{c}\sum_{n} \frac{1}{n!c^{2n}} T_{lm}^{(n)(l+2n+2)}\left(0, t - \frac{r}{c}\right) \Bigg|^2 + \sum_{n} \frac{1}{n!c^{2n}} M_{lm}^{(n)(l+2n+2)}\left(0, t - \frac{r}{c}\right) \Bigg|^2 \Bigg\} \tag{4.169}$$

Now, let us consider the equation above for a dipole case ($l = 1$, $n = n' = 0$). Thereby, the radiation intensity in the case of the electric, magnetic, and toroidal dipole can be approximated as following:

$$I_{rad} = \frac{2}{3c^3} \ddot{\vec{d}}^2 \tag{4.170}$$

$$I_{rad} = \frac{2}{3c^3} \ddot{\vec{m}}^2 \tag{4.171}$$

and

$$I_{rad} = \frac{2}{3c^5} \dddot{\vec{t}}^2 \tag{4.172}$$

respectively. However, it should be noted that to the order of $1/c^5$, there are other contributions too, and straightforwardly, the appropriate formula for the radiation intensity valid to the order of $1/c^5$ can be derived using (4.198). This method yields an equation in terms of the lower electric, magnetic, and toroidal multipole quantities:

$$I_{rad} = \frac{2}{3c^3} \ddot{\vec{d}}^2 + \frac{2}{3c^3} \ddot{\vec{m}}^2 - \frac{4}{3c^4} \ddot{\vec{d}} \cdot \dddot{\vec{t}} + \frac{2}{3c^5} \dddot{\vec{t}}^2 + \frac{2}{15c} \ddot{\vec{m}} \cdot \dddot{\vec{\rho}}^2 + \frac{1}{5c^5} \dddot{Q}_{\alpha\beta}^2 + \frac{1}{20c^5} \dddot{m}_{\alpha\beta}^2 \tag{4.173}$$

where $\vec{d}$, $\vec{m}$, and $\vec{t}$ are the electric, magnetic, and toroidal dipoles, respectively.$\vec{\rho}^2$ is the first mean-square radius of the magnetic dipole distribution, and $Q_{\alpha\beta}$ and $m_{\alpha\beta}$ are the electric and magnetic quadrupole moments, respectively, in which can be extended as:

$$\dddot{Q}_{\alpha\beta}^2 \equiv \dddot{Q}_{\alpha\beta} \dddot{Q}_{\alpha\beta} = \dddot{Q}_{xx}^2 + \dddot{Q}_{yy}^2 + \dddot{Q}_{zz}^2 + 2\left(\dddot{Q}_{xy}^2 + \dddot{Q}_{yz}^2 + \dddot{Q}_{yz}^2\right) \tag{4.174}$$

$$\dddot{m}_{\alpha\beta}^2 \equiv \dddot{m}_{\alpha\beta}\dddot{m}_{\alpha\beta} = \dddot{m}_{xx}^2 + \dddot{m}_{yy}^2 + \dddot{m}_{zz}^2 + 2\left(\dddot{m}_{xy}^2 + \dddot{m}_{yz}^2 + \dddot{m}_{yz}^2\right) \tag{4.175}$$

where the following relations were employed:

$$Q_{xx} + Q_{yy} + Q_{zz} = 0,\ Q_{ij} = Q_{ji} \tag{4.176}$$

$$m_{xx} + m_{yy} + m_{zz} = 0,\ m_{ij} = m_{ji} \tag{4.177}$$

4.3 Angular Momentum Loss

To study the angular momentum loss in a particular system, similar to previous analyses, we initially consider an arbitrary system of charges and currents expressed in terms of charge and current densities that satisfy the following continuity relation:

$$\frac{\partial \rho(\vec{r}, t)}{\partial t} + \text{div}\, \vec{j}(\vec{r}, t) = 0 \tag{4.178}$$

Using the electric and magnetic fields of the source (i.e.,$\vec{E}(\vec{r}, t)$ and $\vec{B}(\vec{r}, t)$) at large distances, the angular momentum loss induced by the source per unit time can be defined by Landau and Lifshitz (1993):

$$\frac{d\vec{M}}{dt} = \lim_{r\to\infty} \frac{r^3}{4\pi} \int d\Omega \left[(\vec{n}\cdot\vec{E})(\vec{n}\times\vec{E}) + (\vec{n}\cdot\vec{B})(\vec{n}\times\vec{B})\right] \tag{4.179}$$

In (4.179), the total angular momentum loss by the system per unit time is only the flux of angular momentum of the radiation field through a spherical surface of large radius ($r \to \infty$):

$$\frac{dM_i}{dt} = \int \varepsilon_{ijk} x_j \sigma_{kl} n_l dS, \quad dS = r^2 d\Omega, \quad n = \frac{\vec{r}}{r} \tag{4.180}$$

where σ_{kl} is the three-dimensional Maxwell stress tensor, as:

$$\sigma_{ij} = \frac{1}{4\pi}\left[E_i E_j + B_i B_j - \frac{1}{2}\delta_{ij}(E^2 + B^2)\right] \tag{4.181}$$

Although (4.179) cannot be applied to the radiation fields at large distances, the fields can be defined to the order of $1/r$. In this limit, since $\vec{n}\cdot\vec{E} = \vec{n}\cdot\vec{B} = 0$, the integrand vanishes. On the other hand, the fields $\vec{E}$ and $\vec{B}$ to the order of $O(1/r)$ can be utilized to get the factors $\vec{n}\times\vec{E}$ and $\vec{n}\times\vec{B}$. Here, the longitudinal components $\vec{n}\cdot\vec{E}$ and $\vec{n}\cdot\vec{B}$ are appeared due to the contributions of the next order $O(1/r^2)$, and the integral in (4.179) is on the order of $O(1/r^3)$, thus, the distance r essentially disappears from the results. Here, it is important to remind that the expressions of $\vec{n} \times \vec{E}$ and $\vec{n} \times \vec{B}$ to the order $O(1/r)$ and the definitions of $\vec{n} \cdot \vec{E}$ and $\vec{n} \cdot \vec{B}$ to the order $O(1/r^2)$ were discussed in previous sections.

The contributions of the electric and magnetic fields to the angular momentum loss can be written by considering (4.179):

$$\frac{d\vec{M}}{dt} = \frac{d\vec{M}^{(el)}}{dt} + \frac{d\vec{M}^{(mag)}}{dt} \tag{4.182}$$

where

$$\frac{d\vec{M}^{(el)}}{dt} = \lim_{r \to \infty} \frac{r^3}{4\pi} \int d\Omega [(\vec{n} \cdot \vec{E})(\vec{n} \times \vec{E})] \tag{4.183}$$

$$\frac{d\vec{M}^{(mag)}}{dt} = \lim_{r \to \infty} \frac{r^3}{4\pi} \int d\Omega [(\vec{n} \cdot \vec{B})(\vec{n} \times \vec{B})] \tag{4.184}$$

To investigate the contribution of the electric field to the angular momentum loss, one should recall (4.147) and (4.159), then the following equation should be calculated:

$$\frac{d\vec{M}^{(el)}}{dt} = \lim_{r \to \infty} \frac{r^3}{4\pi} \left(\vec{J}_- + \vec{J}_+ + \vec{J}_0 \right) \tag{4.185}$$

where

$$\begin{aligned}
\vec{J}_- = \frac{2\sqrt{\pi}}{r^2} & \left\{ \sum_{l,m} \frac{1}{c^l} \frac{(l+1)\sqrt{2l+1}}{(2l+1)!!} \left[Q_{lm}^{(0)(l)}\left(0, t - \frac{r}{c}\right) \right.\right. \\
& \left.\left. - \sum_n \frac{1}{n! c^{2n+1}} T_{lm}^{(n)(l+2n+1)}\left(0, t - \frac{r}{c}\right) \right] \right\} \frac{2\sqrt{\pi}}{r} \\
& \times \left\{ \sum_{l'm'n'} \frac{1}{n'! c^{l'+2n'+1}} \frac{l'+1}{\sqrt{l'}(2l'+1)!!} M_{l'm'}^{(n')(l'+2n'+1)}\left(0, t - \frac{r}{c}\right) \right\} \\
& \times \int d\Omega Y_{lm}(\vec{n}) Y_{l'l'-1m'}(\vec{n})
\end{aligned} \tag{4.186}$$

$$\begin{aligned}
\vec{J}_+ = \frac{2\sqrt{\pi}}{r^2} & \left\{ \sum_{l,m} \frac{1}{c^l} \frac{(l+1)\sqrt{2l+1}}{(2l+1)!!} \left[Q_{lm}^{(0)(l)}\left(0, t - \frac{r}{c}\right) \right.\right. \\
& \left.\left. - \sum_n \frac{1}{n! c^{2n+1}} T_{lm}^{(n)(l+2n+1)}\left(0, t - \frac{r}{c}\right) \right] \right\} \frac{2\sqrt{\pi}}{r}
\end{aligned}$$

$$\times \left\{ \sum_{l'm'n'} \frac{1}{n'!c^{l'+2n'+1}} \frac{l'+1}{\sqrt{l'}(2l'+1)!!} M_{l'm'}^{(n')(l'+2n'+1)}\left(0, t-\frac{r}{c}\right) \right\}$$
$$\times \int d\Omega Y_{lm}(\vec{n}) Y_{l'l'+1m'}(\vec{n}) \tag{4.187}$$

$$\vec{J}_0 = \frac{2\sqrt{\pi}}{r^2} \left\{ \sum_{l,m} \frac{1}{c^l} \frac{(l+1)\sqrt{2l+1}}{(2l+1)!!} \left[Q_{lm}^{(0)(l)}\left(0, t-\frac{r}{c}\right) \right.\right.$$
$$\left.\left. - \sum_n \frac{1}{n!c^{2n+1}} T_{lm}^{(n)(l+2n+1)}\left(0, t-\frac{r}{c}\right) \right] \right\} \frac{2i\sqrt{\pi}}{r}$$
$$\times \left\{ \sum_{l'm'n'} \frac{1}{c^{l'+1}} \frac{\sqrt{(2l'+1)(l'+1)}}{\sqrt{l'}(2l'+1)!!} \left[-Q_{l'm'}^{(0)(l'+1)}\left(0, t-\frac{r}{c}\right) \right.\right.$$
$$\left.\left. + \sum_{n'} \frac{1}{n'!c^{2n'+1}} T_{lm}^{(n')(l'+2n'+1)}\left(0, t-\frac{r}{c}\right) \right] \right\} \int d\Omega Y_{lm}(\vec{n}) Y_{l'l'm'}(\vec{n}) \tag{4.188}$$

Furthermore, the integrals across a solid angle still remains in (4.186)–(4.188), which can be easily assessed via the Cartesian components of the spherical vectors $\vec{Y}_{ll'm}(l' = l, l' = l \pm 1)$ using conventional normalization:

$$\int d\Omega Y_{lm}(\vec{n}) Y^*_{l'm'}(\vec{n}) = \delta_{ll'} \delta_{mm'} \tag{4.189}$$

$$\int d\Omega Y_{lm}(\vec{n}) Y_{l'm'}(\vec{n}) = (-1)^m \delta_{ll'} \delta_{m,-m'} \tag{4.190}$$

where these integrals can be expanded as:

$$\int d\Omega Y_{lm}(\vec{n}) Y_{l'm'}(\vec{n})$$
$$= \vec{e}_x \left[\frac{(-1)^m}{2} \sqrt{\frac{(l+m)(l-m+1)}{l(l+1)}} \delta_{ll'} \delta_{m,-m'+1} \right.$$
$$\left. + \frac{(-1)^m}{2} \sqrt{\frac{(l-m)(l+m+1)}{l(l+1)}} \delta_{ll'} \delta_{m,-m'-1} \right]$$
$$+ \vec{e}_y \left[\frac{i(-1)^m}{2} \sqrt{\frac{(l+m)(l-m+1)}{l(l+1)}} \delta_{ll'} \delta_{m,-m'+1} \right.$$
$$\left. - \frac{(-1)^m}{2} \sqrt{\frac{(l-m)(l+m+1)}{l(l+1)}} \delta_{ll'} \delta_{m,-m'-1} \right]$$

$$+\vec{e}_z\left[(-1)^{m+1}\frac{m}{\sqrt{l(l+1)}}\delta_{ll'}\delta_{m,-m'}\right] \tag{4.191}$$

$$\begin{aligned}
&\int d\Omega Y_{lm}(\vec{n})Y_{l'l'-1m'}(\vec{n})\\
&=\vec{e}_x\left[-\frac{(-1)^m}{2}\sqrt{\frac{(l-m+1)(l-m+2)}{(l+1)(2l+1)}}\delta_{l,l'-1}\delta_{m,-m'+1}\right.\\
&\left.+\frac{(-1)^m}{2}\sqrt{\frac{(l+m+1)(l+m+2)}{(l+1)(2l+1)}}\delta_{l,l'-1}\delta_{m,-m'-1}\right]\\
&+\vec{e}_y\left[-\frac{i(-1)^m}{2}\sqrt{\frac{(l-m+1)(l-m+2)}{(l+1)(2l+1)}}\delta_{l,l'-1}\delta_{m,-m'+1}\right.\\
&\left.-\frac{i(-1)^m}{2}\sqrt{\frac{(l+m+1)(l+m+2)}{(l+1)(2l+1)}}\delta_{l,l'-1}\delta_{m,-m'-1}\right]\\
&+\vec{e}_z\left[(-1)^m\sqrt{\frac{(l+m+1)(l-m+1)}{(l+1)(2l+1)}}\delta_{l,l'-1}\delta_{m,-m'}\right]
\end{aligned} \tag{4.192}$$

$$\begin{aligned}
&\int d\Omega Y_{lm}(\vec{n})Y_{l'l'+1m'}(\vec{n})\\
&=\vec{e}_x\left[-\frac{(-1)^m}{2}\sqrt{\frac{(l+m-1)(l+m)}{l(2l+1)}}\delta_{l,l'+1}\delta_{m,-m'+1}\right.\\
&\left.+\frac{(-1)^m}{2}\sqrt{\frac{(l-m-1)(l-m)}{l(2l+1)}}\delta_{l,l'-1}\delta_{m,-m'-1}\right]\\
&+\vec{e}_y\left[-\frac{i(-1)^m}{2}\sqrt{\frac{(l+m-1)(l+m)}{l(2l+1)}}\delta_{l,l'+1}\delta_{m,-m'+1}\right.\\
&\left.-\frac{i(-1)^m}{2}\sqrt{\frac{(l-m-1)(l-m)}{l(2l+1)}}\delta_{l,l'+1}\delta_{m,-m'-1}\right]\\
&+\vec{e}_z\left[-(-1)^m\sqrt{\frac{(l-m)(l+m)}{l(2l+1)}}\delta_{l,l'+1}\delta_{m,-m'}\right]
\end{aligned} \tag{4.193}$$

Using this approach, the contribution of the electric field to the rate of angular momentum loss can be determined. By separately rewriting the contributions of $\vec{J}_-$, $\vec{J}_+$, $\vec{J}_0$ to $d\vec{M}^{(el)}/dt$, we have:

$$\frac{d\vec{M}^{(el)}}{dt}=\left(\frac{d\vec{M}^{(el)}}{dt}\right)^{(\vec{J}_-)}+\left(\frac{d\vec{M}^{(el)}}{dt}\right)^{(\vec{J}_+)}+\left(\frac{d\vec{M}^{(el)}}{dt}\right)^{(\vec{J}_0)}$$

$$\left(\frac{d\vec{M}^{(el)}}{dt}\right)^{(\vec{J}_a)} = \lim_{r\to\infty}\frac{r^3}{4\pi}\vec{J}_a,\ (a) = -, +, 0 \tag{4.194}$$

Thus, we can write:

$$\left(\frac{d\vec{M}^{(el)}}{dt}\right)_x^{(\vec{J}_-)} = \frac{1}{2}\sum_{l,m,n'}\frac{(-1)^{m+1}(l+2)}{(2l+1)!!(2l+3)!!}\frac{1}{n'!c^{2l+2n'+2}}\Bigg[Q_{lm}^{(0)(l)}\left(0, t-\frac{r}{c}\right)$$
$$-\sum_n\frac{1}{n'!c^{2n+1}}T_{lm}^{(n)(l+2n+1)}\left(0, t-\frac{r}{c}\right)\Bigg]\Bigg[\sqrt{(l-m+1)(l-m+2)}$$
$$\times M_{l+1,1-m}^{(n')(l'+2n'+2)}\left(0, t-\frac{r}{c}\right) - \sqrt{(l-m+1)(l-m+2)}$$
$$\times M_{l+1,-1-m}^{(n')(l+2n'+2)}\left(0, t-\frac{r}{c}\right)\Bigg] \tag{4.195}$$

$$\left(\frac{d\vec{M}^{(el)}}{dt}\right)_y^{(\vec{J}_-)} = \frac{1}{2}\sum_{l,m,n'}\frac{i(-1)^{m+1}(l+2)}{(2l+1)!!(2l+3)!!}\frac{1}{n'!c^{2l+2n'+2}}\Bigg[Q_{lm}^{(0)(l)}\left(0, t-\frac{r}{c}\right)$$
$$-\sum_n\frac{1}{n'!c^{2n+1}}T_{lm}^{(n)(l+2n+1)}\left(0, t-\frac{r}{c}\right)\Bigg]\Bigg[\sqrt{(l-m+1)(l-m+2)}$$
$$\times M_{l+1,1-m}^{(n')(l'+2n'+2)}\left(0, t-\frac{r}{c}\right) + \sqrt{(l+m+1)(l+m+2)}$$
$$\times M_{l+1,-1-m}^{(n')(l+2n'+2)}\left(0, t-\frac{r}{c}\right)\Bigg] \tag{4.196}$$

$$\left(\frac{d\vec{M}^{(el)}}{dt}\right)_z^{(\vec{J}_-)} = \sum_{l,m,n'}\frac{(-1)^{m}(l+2)}{(2l+1)!!(2l+3)!!}\frac{\sqrt{(l+m+1)(l+m+2)}}{n'!c^{2l+2n'+2}}$$
$$\times\Bigg[Q_{lm}^{(0)(l)}\left(0, t-\frac{r}{c}\right) - \sum_n\frac{1}{n'!c^{2n+1}}T_{lm}^{(n)(l+2n+1)}\left(0, t-\frac{r}{c}\right)\Bigg]$$
$$\times M_{l+1,-m}^{(n')(l+2n'+2)}\left(0, t-\frac{r}{c}\right)\Bigg] \tag{4.197}$$

$$\left(\frac{d\vec{M}^{(el)}}{dt}\right)_x^{(\vec{J}_+)} = \frac{1}{2}\sum_{l,m,n'}\frac{(-1)^{m+1}(l+1)}{(2l-1)!!(2l+1)!!}\frac{1}{n'!c^{2l+2n'}}\Bigg[Q_{lm}^{(0)(l)}\left(0, t-\frac{r}{c}\right)$$
$$-\sum_n\frac{1}{n!c^{2n+1}}T_{lm}^{(n)(l+2n+1)}\left(0, t-\frac{r}{c}\right)\Bigg]\Bigg[\sqrt{(l+m-1)(l+m)}$$
$$\times M_{l-1,1-m}^{(n')(l'+2n')}\left(0, t-\frac{r}{c}\right) - \sqrt{(l-m-1)(l-m)}$$
$$\times M_{l-1,-1-m}^{(n')(l+2n')}\left(0, t-\frac{r}{c}\right)\Bigg] \tag{4.198}$$

$$\left(\frac{d\vec{M}^{(el)}}{dt}\right)_y^{(\vec{J}_+)} = \frac{1}{2}\sum_{l,m,n'}\frac{i(-1)^{m+1}(l+1)}{(2l-1)!!(2l+1)!!}\frac{1}{n'!c^{2l+2n'}}\left[Q_{lm}^{(0)(l)}\left(0,t-\frac{r}{c}\right)\right.$$
$$\left.-\sum_n \frac{1}{n!c^{2n+1}}T_{lm}^{(n)(l+2n+1)}\left(0,t-\frac{r}{c}\right)\right]\left[\sqrt{(l+m-1)(l+m)}\right.$$
$$\times M_{l-1,1-m}^{(n')(l'+2n')}\left(0,t-\frac{r}{c}\right)+\sqrt{(l-m-1)(l-m)}$$
$$\left.\times M_{l-1,-1-m}^{(n')(l+2n')}\left(0,t-\frac{r}{c}\right)\right] \quad (4.199)$$

$$\left(\frac{d\vec{M}^{(el)}}{dt}\right)_z^{(\vec{J}_+)} = \sum_{l,m,n'}\frac{(-1)^{m+1}(l+1)}{(2l-1)!!(2l+1)!!}\frac{\sqrt{(l-m)(l+m)}}{n'!c^{2l+2n'}}\left[Q_{lm}^{(0)(l)}\left(0,t-\frac{r}{c}\right)\right.$$
$$\left.-\sum_n \frac{1}{n!c^{2n+1}}T_{lm}^{(n)(l+2n+1)}\left(0,t-\frac{r}{c}\right)\right]M_{l-1,-m}^{(n')(l+2n')}\left(0,t-\frac{r}{c}\right)$$
$$(4.200)$$

$$\left(\frac{d\vec{M}^{(el)}}{dt}\right)_x^{(\vec{J}_0)} = \frac{i}{2}\sum_{l,m}\frac{(-1)^{m+1}(l+1)}{l(2l-1)!!(2l+1)!!}\frac{1}{n'!c^{2l+1}}\left[Q_{lm}^{(0)(l)}\left(0,t-\frac{r}{c}\right)\right.$$
$$\left.-\sum_n \frac{1}{n!c^{2n+1}}T_{lm}^{(n)(l+2n+1)}\left(0,t-\frac{r}{c}\right)\right]\left\{\sqrt{(l-m+1)(l+m)}\right.$$
$$\times\left[Q_{l,1-m}^{(0)(l+1)}\left(0,t-\frac{r}{c}\right)+\sum_{n'}\frac{1}{n'!c^{2n'+1}}T_{l,1-m}^{(n')(l'+2n'+2)}\left(0,t-\frac{r}{c}\right)\right]$$
$$+\sqrt{(l+m+1)(l-m)}\left[-Q_{l,-1-m}^{(0)(l+1)}\left(0,t-\frac{r}{c}\right)\right.$$
$$\left.\left.+\sum_{n'}\frac{1}{n'!c^{2n'+1}}T_{l,-1-m}^{(n')(l'+2n'+2)}\left(0,t-\frac{r}{c}\right)\right]\right\} \quad (4.201)$$

$$\left(\frac{d\vec{M}^{(el)}}{dt}\right)_y^{(\vec{J}_0)} = \frac{1}{2}\sum_{l,m}\frac{(-1)^{m}(l+1)}{l(2l-1)!!(2l+1)!!}\frac{1}{c^{2l+1}}\left[Q_{lm}^{(0)(l)}\left(0,t-\frac{r}{c}\right)\right.$$
$$\left.-\sum_n \frac{1}{n!c^{2n+1}}T_{lm}^{(n)(l+2n+1)}\left(0,t-\frac{r}{c}\right)\right]\left\{\sqrt{(l-m+1)(l+m)}\right.$$
$$\times\left[Q_{l,1-m}^{(0)(l+1)}\left(0,t-\frac{r}{c}\right)+\sum_{n'}\frac{1}{n'!c^{2n'+1}}T_{l,1-m}^{(n')(l'+2n'+2)}\left(0,t-\frac{r}{c}\right)\right]$$
$$-\sqrt{(l+m+1)(l-m)}\left[-Q_{l,-1-m}^{(0)(l+1)}\left(0,t-\frac{r}{c}\right)\right.$$

$$+\sum_{n'}\frac{1}{n'!c^{2n'+1}}T_{l,-m}^{(n')(l'+2n'+2)}\left(0,t-\frac{r}{c}\right)\Bigg]\Bigg\} \tag{4.202}$$

$$\begin{aligned}\left(\frac{d\vec{M}^{(el)}}{dt}\right)_z^{(\vec{J}_0)} &= \sum_{l,m}\frac{i(-1)^m(l+1)}{l(2l-1)!!(2l+1)!!}\frac{1}{c^{2l+1}}\Bigg[Q_{lm}^{(0)(l)}\left(0,t-\frac{r}{c}\right)\\ &-\sum_{n}\frac{1}{n!c^{2n+1}}T_{lm}^{(n)(l+2n+1)}\left(0,t-\frac{r}{c}\right)\Bigg]\Bigg[-Q_{l,-m}^{(0)(l+1)}\left(0,t-\frac{r}{c}\right)\\ &+\sum_{n'}\frac{1}{n'!c^{2n'+1}}T_{l,-m}^{(n')(l'+2n'+2)}\left(0,t-\frac{r}{c}\right)\Bigg]\end{aligned} \tag{4.203}$$

In the set of equations above, the magnetic multipole moments and radii of the system contribute to $d\vec{M}^{(el)}/dt$ through the interferences between the electric and toroidal components. In other words, they only contribute through $\vec{J}_{\pm}$ parts in [(4.195)–(4.200)].

The calculation of the magnetic field contribution to the angular momentum loss in (4.184) is similar to the electric field derivation. However, it should be noted that this detailed expression can be employed in a slightly different way as we did before, where various types of interferences must be considered during the calculations. Consequently, one has:

$$\frac{d\vec{M}^{(mag)}}{dt}=\left(\frac{d\vec{M}^{(mag)}}{dt}\right)_{M,M}+\left(\frac{d\vec{M}^{(mag)}}{dt}\right)_{M,Q+T} \tag{4.204}$$

where *M, M* indicates the interferences between the induced magnetic multipole radii, and *M, Q + T* specifies the interferences between the magnetic and electric + toroidal type radii:

$$\begin{aligned}\left(\frac{d\vec{M}_x^{(mag)}}{dt}\right)_{M,M} &= -\frac{i}{2}\sum_{l,m,n,n'}\frac{(-1)^m(l+1)}{l(2l-1)!!(2l+1)!!}\frac{\sqrt{(l+m)(l-m+1)}}{n!n'!c^{2(l+n+n')+1}}\\ &\times M_{lm}^{(n)(l+2n)}\left(0,t-\frac{r}{c}\right)M_{l,1-m}^{(n')(l+2n'+1)}\left(0,t-\frac{r}{c}\right)\\ &-\frac{i}{2}\sum_{l,m,n,n'}\frac{(-1)^m(l+1)}{l(2l-1)!!(2l+1)!!}\frac{\sqrt{(l+m)(l+m+1)}}{n!n'!c^{2(l+n+n')+1}}\\ &\times M_{lm}^{(n)(l+2n)}\left(0,t-\frac{r}{c}\right)M_{l,-1-m}^{(n')(l+2n'+1)}\left(0,t-\frac{r}{c}\right)\end{aligned} \tag{4.205}$$

$$\begin{aligned}\left(\frac{d\vec{M}_y^{(mag)}}{dt}\right)_{M,M} &= \frac{1}{2}\sum_{l,m,n,n'}\frac{(-1)^m(l+1)}{l(2l-1)!!(2l+1)!!}\frac{\sqrt{(l+m)(l-m+1)}}{n!n'!c^{2(l+n+n')+1}}\\ &\times M_{lm}^{(n)(l+2n)}\left(0,t-\frac{r}{c}\right)M_{l,1-m}^{(n')(l+2n'+1)}\left(0,t-\frac{r}{c}\right)\end{aligned}$$

$$-\frac{1}{2}\sum_{l,m,n,n'}\frac{(-1)^m(l+1)}{l(2l-1)!!(2l+1)!!}\frac{\sqrt{(l+m)(l+m+1)}}{n!n'!c^{2(l+n+n')+1}}$$
$$\times M_{lm}^{(n)(l+2n)}\left(0,t-\frac{r}{c}\right)M_{l,-1-m}^{(n')(l+2n'+1)}\left(0,t-\frac{r}{c}\right) \qquad (4.206)$$

$$\left(\frac{d\vec{M}_z^{(mag)}}{dt}\right)_{M,M}=i\sum_{l,m,n,n'}\frac{(-1)^m(l+1)}{l(2l-1)!!(2l+1)!!}\frac{1}{n!n'!c^{2(l+n+n')+1}}$$
$$\times M_{lm}^{(n)(l+2n)}\left(0,t-\frac{r}{c}\right)M_{l,-m}^{(n')(l+2n'+1)}\left(0,t-\frac{r}{c}\right) \qquad (4.207)$$

$$\left(\frac{d\vec{M}_x^{(mag)}}{dt}\right)_{M,Q+T}=\frac{1}{2}\sum_{l,m,n}\frac{(-1)^m}{(2l-1)!!(2l+1)!!}\frac{1}{n!c^{l+2n}}M_{lm}^{(n)(l+2n)}\left(0,t-\frac{r}{c}\right)$$
$$\times\left\{\frac{(l+2)\sqrt{(l-m+1)(l-m+2)}}{(2l+1)(2l+3)}\right.$$
$$\times\left[\frac{Q_{l+1,1-m}^{(0)(l+2n)}\left(0,t-\frac{r}{c}\right)}{c^{l+2}}-\sum_{n'=0}^{\infty}\frac{T_{l+1,1-m}^{(n')(l+2n'+3)}\left(0,t-\frac{r}{c}\right)}{n!c^{l+2n'+3}}\right]$$
$$-\frac{(l+2)\sqrt{(l+m+1)(l+m+2)}}{(2l+1)(2l+3)}$$
$$\times\left[\frac{Q_{l+1,1-m}^{(0)(l+2n)}\left(0,t-\frac{r}{c}\right)}{c^{l+2}}-\sum_{n'=0}^{\infty}\frac{T_{l+1,-1-m}^{(n')(l+2n'+3)}\left(0,t-\frac{r}{c}\right)}{n'!c^{l+2n'+3}}\right]$$
$$+(l+1)\sqrt{(l+m-1)(l+m)}$$
$$\times\left[\frac{Q_{l-1,-1-m}^{(0)(l+2n)}\left(0,t-\frac{r}{c}\right)}{c^{l}}-\sum_{n'=0}^{\infty}\frac{T_{l-1,1-m}^{(n')(l+2n'+1)}\left(0,t-\frac{r}{c}\right)}{n'!c^{l+2n'+1}}\right]$$
$$-(l+1)\sqrt{(l-m-1)(l-m)}$$
$$\left.\times\left[\frac{Q_{l-1,-1-m}^{(0)(l)}\left(0,t-\frac{r}{c}\right)}{c^{l}}-\sum_{n'=0}^{\infty}\frac{T_{l-1,-1-m}^{(n')(l+2n'+1)}\left(0,t-\frac{r}{c}\right)}{n'!c^{l+2n'+1}}\right]\right\}$$
$$(4.208)$$

$$\left(\frac{d\vec{M}_y^{(mag)}}{dt}\right)_{M,Q+T}=\frac{i}{2}\sum_{l,m,n}\frac{(-1)^m}{(2l-1)!!(2l+1)!!}\frac{1}{n!c^{l+2n}}M_{lm}^{(n)(l+2n)}\left(0,t-\frac{r}{c}\right)$$
$$\times\left\{\frac{(l+2)\sqrt{(l-m+1)(l-m+2)}}{(2l+1)(2l+3)}\right.$$
$$\times\left[\frac{Q_{l+1,1-m}^{(0)(l+2n)}\left(0,t-\frac{r}{c}\right)}{c^{l+2}}-\sum_{n'=0}^{\infty}\frac{T_{l+1,1-m}^{(n')(l+2n'+3)}\left(0,t-\frac{r}{c}\right)}{n!c^{l+2n'+3}}\right]$$
$$-\frac{(l+2)\sqrt{(l+m+1)(l+m+2)}}{(2l+1)(2l+3)}$$

$$\times \left[\frac{Q^{(0)(l+2n)}_{l+1,-1-m}\left(0, t - \frac{r}{c}\right)}{c^{l+2}} - \sum_{n'=0}^{\infty} \frac{T^{(n')(l+2n'+3)}_{l+1,-1-m}\left(0, t - \frac{r}{c}\right)}{n'!c^{l+2n'+3}} \right]$$
$$+ (l+1)\sqrt{(l+m-1)(l+m)}$$
$$\times \left[\frac{Q^{(0)(l+2n)}_{l-1,1-m}\left(0, t - \frac{r}{c}\right)}{c^{l}} - \sum_{n'=0}^{\infty} \frac{T^{(n')(l+2n'+1)}_{l-1,1-m}\left(0, t - \frac{r}{c}\right)}{n'!c^{l+2n'+1}} \right]$$
$$+ (l+1)\sqrt{(l-m-1)(l-m)}$$
$$\times \left[\frac{Q^{(0)(l)}_{l-1,-1-m}\left(0, t - \frac{r}{c}\right)}{c^{l}} - \sum_{n'=0}^{\infty} \frac{T^{(n')(l+2n'+1)}_{l-1,-1-m}\left(0, t - \frac{r}{c}\right)}{n'!c^{l+2n'+1}} \right]\Bigg\} \tag{4.209}$$

$$\left(\frac{d\vec{M}_z^{(mag)}}{dt}\right)_{M,Q+T} = \sum_{l,m,n} \frac{(-1)^m}{(2l-1)!!(2l+1)!!} \frac{1}{n!c^{l+2n}} M^{(n)(l+2n)}_{lm}\left(0, t - \frac{r}{c}\right)$$
$$\times \Bigg\{ - \frac{(l+2)\sqrt{(l+m+1)(l-m+1)}}{(2l+1)(2l+3)}$$
$$\times \left[\frac{Q^{(0)(l+2n)}_{l+1,-m}\left(0, t - \frac{r}{c}\right)}{c^{l+2}} - \sum_{n'=0}^{\infty} \frac{T^{(n')(l+2n'+3)}_{l+1,-m}\left(0, t - \frac{r}{c}\right)}{n!c^{l+2n'+3}} \right]$$
$$+ (l+1)\sqrt{(l-m)(l+m)}$$
$$\times \left[\frac{Q^{(0)(l+2n)}_{l-1,1-m}\left(0, t - \frac{r}{c}\right)}{c^{l}} - \sum_{n'=0}^{\infty} \frac{T^{(n')(l+2n'+1)}_{l-1,-m}\left(0, t - \frac{r}{c}\right)}{n'!c^{l+2n'+1}} \right]\Bigg\} \tag{4.210}$$

By far, the calculation of angular momentum loss per unit time is presented. The electric field contribution in (4.182) is specified in (4.195)–(4.203), while the contribution of the magnetic field in (4.182) is stated in (4.205)–(4.210). The results above were explained by providing formulas for the Cartesian components of $d\vec{M}/dt$. However, these results can be arranged and presented in a much more compact form (known as *Clebsch-Gordan coefficients*) through rewriting them in terms of spherical components ($\mu = -1,0,1$) of vector $d\vec{M}/dt$. Accordingly, one can define the expression of angular momentum loss per unit volume as:

$$\frac{dM_\mu}{dt} = \frac{(-1)^m}{(2l-1)!!(2l+1)!!} \frac{1}{c^{2l}} \left[\frac{l+2}{2l+1} \frac{\sqrt{l+1}}{\sqrt{2l+3}} C^{l+1,1,l}_{m+\mu,-\mu,m} A \right.$$
$$\left. + (l+1)\sqrt{l(2l-1)} C^{l-1,1,l}_{m+\mu,-\mu,m} B + \frac{i}{c}(l+1)\frac{\sqrt{l+1}}{\sqrt{l}} C^{l,1,l}_{m+\mu,-\mu,m} C \right] \tag{4.211}$$

where A, B, and C are:

$$A = \sum_{n'} \frac{1}{c^{2(n'+1)} n'!} \left[Q_{lm}^{(0)(l)} M_{l+1,-\mu-m}^{(n')(l+2n'+2)} - M_{lm}^{(n')(l+2n')} Q_{l+1,-\mu-m}^{(0)(l+2)} \right]$$
$$+ \sum_{n,n'} \frac{1}{c^{2n'+2+3} n! n'!} \left[M_{lm}^{(n')(l+2n')} T_{l+1,-\mu-m}^{(n)(l+2n+3)} - T_{lm}^{(n)(l+2n+1)} M_{l+1,-\mu-m}^{(n')(l+2n'+2)} \right] \tag{4.212}$$

$$B = \sum_{n'} \frac{1}{c^{2n'} n'!} \left[Q_{lm}^{(0)(l)} M_{l-1,-\mu-m}^{(n')(l+2n')} - M_{lm}^{(n')(l+2n')} Q_{l-1,-\mu-m}^{(0)(l)} \right]$$
$$+ \sum_{n,n'} \frac{1}{c^{2n'+2+1} n! n'!} \left[M_{lm}^{(n')(l+2n')} T_{l-1,-\mu-m}^{(n)(l+2n+1)} - T_{lm}^{(n)(l+2n+1)} M_{l-1,-\mu-m}^{(n')(l+2n')} \right] \tag{4.213}$$

$$C = -Q_{lm}^{(0)(l)} Q_{l,-\mu-m}^{(0)(l+1)} + \sum_{n} \frac{1}{c^{2n+1} n!} \left[Q_{lm}^{(0)(l)} T_{l,-\mu-m}^{(n)(l+2n+2)} - T_{lm}^{(n)(l+2n+1)} Q_{l,-\mu-m}^{(0)(l+1)} \right]$$
$$- \sum_{n,n'} \frac{1}{c^{2(n'+1)} n! n'!} \left[M_{lm}^{(n)(l+2n')} M_{l,-\mu-m}^{(n')(l+2n'+1)} + \frac{1}{c^2} T_{lm}^{(n)(l+2n+1)} M_{l,-\mu-m}^{(n')(l+2n'+2)} \right] \tag{4.214}$$

Besides, the connection with the Cartesian components can be defined as:

$$\begin{aligned} \frac{dM_+}{dt} &= -\frac{1}{\sqrt{2}} \left(\frac{dM_x}{dt} + i \frac{dM_y}{dt} \right) \\ \frac{dM_-}{dt} &= -\frac{1}{\sqrt{2}} \left(\frac{dM_x}{dt} - i \frac{dM_y}{dt} \right) \\ \frac{dM_{(0)}}{dt} &= \frac{dM_z}{dt} \end{aligned} \tag{4.215}$$

Considering previously used double superscript quantities with the argument of $t-r/c$ ($Q_{lm}^{(n)(\upsilon)}$, $M_{lm}^{(n)(\upsilon)}$, $T_{lm}^{(n)(\upsilon)}$), now, one can recall the definitions of double-superscript quantities in terms of the corresponding multipole electric, magnetic, and toroidal form factors in (4.13)–(4.15). Also, by utilizing their relation with the multipole electric ($\bar{r}_{lm}^{2n}(t)$), magnetic ($\bar{\rho}_{lm}^{2n}(t)$), and toroidal ($\bar{R}_{lm}^{2n}(t)$) mean-square radii of various orders, one has:

$$Q_{lm}^{(n)(\upsilon)}(0,t) = \frac{(2l+1)!!}{2^n (2l+2n+1)!!} \frac{d^\upsilon}{dt^\upsilon} \bar{r}_{lm}^{2n}(t) \tag{4.216}$$

$$M_{lm}^{(n)(\upsilon)}(0,t) = \frac{(2l+1)!!}{2^n (2l+2n+1)!!} \frac{d^\upsilon}{dt^\upsilon} \bar{\rho}_{lm}^{2n}(t) \tag{4.217}$$

$$T_{lm}^{(n)(\upsilon)}(0,t) = \frac{(2l+1)!!}{2^n (2l+2n+1)!!} \frac{d^\upsilon}{dt^\upsilon} \bar{R}_{lm}^{2n}(t) \tag{4.218}$$

The radii above describe the multipole content of the most general type of source. Here, one should consider the time dependence in the formulas obtained for the angular momentum loss per unit time. This point, indeed, raises an issue: the evaluation of the limits $r \to \infty$ in (4.179) might rise some tricky problems. This is because the electric and magnetic fields in the integral are solutions of the Maxwell equations that are zero at infinity. However, one may calculate the angular momentum loss of a system using a sphere of radius r in (4.179).

Since it is challenging to perform the described process above within the proximity of the source, one could perform calculations for large r, but not in the limit of $r \to \infty$. This method allows for having non-negligible terms of order $1/r^3$ from the integral to compensate the r^3. Therefore, one needs to consider $d\vec{M}(r,t)/dt$ as a function of r. It is important to note that the argument of the radii in the obtained result for $d\vec{M}/dt$ is *t-r/c*. This implies that when an observer at a distance r_1 from the source measures the angular momentum, another observer located at a distance r_2, where $r_2 > r_1$, will obtain similar results as the first observer within the same time interval T. This explicitly describes why $d\vec{M}/dt$ and the recoil force depend on r through the retarded time *t-r/c* in the argument of the radii.

Now, (4.211)–(4.214) can be transcribed correctly by rephrasing the description of the angular momentum loss in terms of $1/c^5$. Using both Cartesian components and the associated spherical components of the first multipole moments and radii, one can write:

$$\frac{dM_\alpha}{dt} = C(QQ) + C(MM) + C(QM) + C(QT) + C(MT) + C(TT) \tag{4.219}$$

where the diagonal and interference contributions of the electric, magnetic, and toroidal components are:

$$C(QQ) = -\frac{2}{3c^3}\varepsilon_{\alpha ij}\dot{d}_i\ddot{d}_j - \frac{2}{5c^5}\varepsilon_{\alpha ij}\ddot{Q}_{\beta i}\dddot{Q}_{\beta j} \tag{4.220}$$

$$\begin{aligned} C(MM) = &-\frac{2}{3c^3}\varepsilon_{\alpha ij}\dot{m}_i\ddot{m}_j - \frac{1}{5c^5}\left(-\frac{1}{3}\varepsilon_{\alpha ij}\ddot{m}_i\dddot{\bar{\rho}}_j^2 + \frac{1}{3}\varepsilon_{\alpha ij}\dot{m}_i\ddddot{\bar{\rho}}_j^2 \right. \\ &\left. + \frac{1}{2}\varepsilon_{\alpha ij}\ddot{m}_{\beta i}\dddot{m}_{\beta j}\right) \end{aligned} \tag{4.221}$$

$$C(MQ) = -\frac{1}{5c^4}\left(\dddot{m}_{\alpha i}\dot{d}_i + \ddot{m}_{\alpha i}\ddot{d}_i - 2\dddot{Q}_{\alpha i}\dot{m}_i - 2\ddot{Q}_{\alpha i}\ddot{m}_i\right) \tag{4.222}$$

$$C(QT) = \frac{2}{3c^4}\varepsilon_{\alpha ij}\left(\dot{d}_i\dddot{t}_j + \ddot{t}_i\ddot{d}_j\right) \tag{4.223}$$

$$C(MT) = -\frac{1}{15c^5}\left(2\ddddot{t}_{\alpha i}\dot{m}_i + 2\dddot{t}_{\alpha i}\ddot{m}_i - 3\dddot{m}_{\alpha i}\ddot{t}_i - 3\ddot{m}_{\alpha i}\dddot{t}_i\right) \tag{4.224}$$

$$C(TT) = -\frac{2}{3c^5}\varepsilon_{\alpha ij}\ddot{t}_i\dddot{t}_j \tag{4.225}$$

In the equations above, dots mean time derivatives, d_i is the electric dipole moment, Q_{ij} is the electric quadrupole moment, m_i is the magnetic dipole moment, $\bar{\rho}_i^2$ is the first mean-square radius of the magnetic dipole distribution, m_{ij} is the magnetic quadrupole moment, t_i is the toroidal dipole moment, and t_{ij} is the toroidal quadrupole moment. It should be noted that the argument of the multipole quantities is *t-r/c*. Considering the above equations to the power of $1/c$, one can obtain:

$$\begin{aligned}\frac{dM_\alpha}{dt} &= -\frac{2}{3c^3}\varepsilon_{\alpha\beta\gamma}\left(\dot{d}_\beta\ddot{d}_\gamma + \dot{m}_\beta\ddot{m}_\gamma\right) + \frac{1}{c^4}\left[\frac{1}{5}\left(-\dddot{m}_{\alpha\beta}\dot{d}_\beta - \ddot{m}_{\alpha\beta}\ddot{d}_\beta\right.\right.\\ &\left.\left.+2\dddot{Q}_{\alpha\beta}\dot{m}_\beta + 2\ddot{Q}_{\alpha\beta}\ddot{m}_\beta\right) + \frac{2}{3}\varepsilon_{\alpha\beta\gamma}\left(\dot{d}_\beta\dddot{t}_\gamma + \ddot{t}_\beta\ddot{d}_\gamma\right)\right] - \frac{1}{5c^5}\\ &\times\left[\frac{1}{3}\left(-2\ddddot{t}_{\alpha\beta}\dot{m}_\beta + 2\dddot{t}_{\alpha\beta}\ddot{m}_\beta - 3\dddot{m}_{\alpha\beta}\ddot{t}_\beta - 3\ddot{m}_{\alpha\beta}\dddot{t}_\beta\right) + \varepsilon_{\alpha\beta\gamma}\right.\\ &\left.\times\left(2\ddot{Q}_{\delta\beta}\dddot{Q}_{\delta\gamma} + \frac{10}{3}\ddot{t}_\beta\dddot{t}_\gamma + \frac{1}{3}\dot{m}_\beta\ddddot{\bar{\rho}}_\gamma^2 - \frac{1}{3}\ddot{m}_\beta\dddot{\bar{\rho}}_\gamma^2 + \frac{1}{2}\ddot{m}_{\delta\beta}\dddot{m}_{\delta\gamma}\right)\right]\\ &+[\mathbb{C}]\end{aligned} \tag{4.226}$$

where $\mathbb{C}$ indicates the terms that are higher in order than $1/c^5$.

4.4 Recoil Force

Principally, the recoil force is the momentum loss by the system per unit time, i.e., the flux of momentum going out from the system through the radiation of electromagnetic waves (Gray 1978):

$$F_i = \int \sigma_{ij}n_j dS, \quad dS = r^2 d\Omega, \quad \vec{n} = \frac{\vec{r}}{r} \tag{4.227}$$

where σ_{ij} is the three-dimensional Maxwell stress tensor, mentioned in (4.181), and the integration must be performed across a sphere of radius r.

The recoil force of a system can be expressed as:

$$\vec{F} = \lim_{r\to\infty}\left\{-\frac{r^2}{4\pi}\int d\Omega\left[\vec{E}(\vec{E}\cdot\vec{n}) + \vec{B}(\vec{B}\cdot\vec{n}) - \vec{n} + \frac{\left(\vec{E}^2 + \vec{B}^2\right)}{2}\right]\right\} \tag{4.228}$$

In this regime, one needs the electric and magnetic fields to the order of $O(1/r)$. At this order, the fields are transversal (i.e.,$\vec{E}\cdot\vec{n} = 0$ and $\vec{B}\cdot\vec{n} = 0$) and the following integrals have to be calculated to define the electric and magnetic fields:

$$\vec{F} = -\frac{r}{8\pi}\int d\Omega\vec{r}\left(\vec{E}^2 + \vec{B}^2\right) \tag{4.229}$$

in which each component can be defined as:

$$\begin{cases} \vec{F}_x = -\frac{r}{8\pi}\int d\Omega x\left(\vec{E}^2 + \vec{B}^2\right) \\ \vec{F}_y = -\frac{r}{8\pi}\int d\Omega y\left(\vec{E}^2 + \vec{B}^2\right) \\ \vec{F}_z = -\frac{r}{8\pi}\int d\omega z\left(\vec{E}^2 + \vec{B}^2\right) \end{cases} \tag{4.230}$$

In the set of integrals above, for $r \to \infty$ with $\vec{E}$ and $\vec{B} \sim O(1/r)$, r can be excluded from the calculations.

To evaluate the recoil force in terms of multipole content of the source, any multipolarity order and an arbitrary time dependence of the quantities should be considered. This implies that the recoil force can be specified based on the time derivatives of the system's various multipole radii. In light of (4.227)–(4.230), one requires only the $O(1/r)$ multipole content of electric and magnetic fields. This can be performed through the use of different steps, as shown below.

First, since the electric and magnetic fields on the order of $O(1/r)$ satisfy:

$$\frac{\vec{r}}{r} \times \vec{E}(\vec{r}, t) \overset{O(1/r)}{=} \vec{B}(\vec{r}, t) \tag{4.231}$$

we can write:

$$\vec{E}^2(\vec{r}, t) = \vec{B}^2(\vec{r}, t) \tag{4.232}$$

Therefore,

$$F_x = -r^2\frac{\sqrt{2}}{4\sqrt{3\pi}}\int d\Omega\left[Y_{1-1}(\vec{n}) - Y_{11}(\vec{n})\right]\vec{B}^2(\vec{r}, t) \tag{4.233}$$

$$F_y = -ir^2\frac{\sqrt{2}}{4\sqrt{3\pi}}\int d\Omega\left[Y_{11}(\vec{n}) - Y_{1-1}(\vec{n})\right]\vec{B}^2(\vec{r}, t) \tag{4.234}$$

$$F_z = -r^2\frac{1}{2\sqrt{3\pi}}\int d\Omega Y_{10}(\vec{n})\vec{B}^2(\vec{r}, t) \tag{4.235}$$

Now, we need to calculate the following surface integrals:

$$I_1 = \int d\Omega Y_{11}(\vec{n})\vec{B}^2(\vec{r}, t) \tag{4.236}$$

$$I_{-1} = \int d\Omega Y_{1-1}(\vec{n})\vec{B}^2(\vec{r}, t) \tag{4.237}$$

$$I_0 = \int d\Omega Y_{10}(\vec{n})\vec{B}^2(\vec{r}, t) \tag{4.238}$$

Based on this information, in the Cartesian basis, the recoil force will be:

$$F_x = -r^2 \frac{\sqrt{2}}{4\sqrt{3\pi}}(I_{-1} - I_1) \tag{4.239}$$

$$F_y = -r^2 \frac{\sqrt{2}}{4\sqrt{3\pi}}(iI_{-1} + iI_1) \tag{4.240}$$

$$F_z = -r^2 \frac{\sqrt{2}}{2\sqrt{3\pi}} I_0 \tag{4.241}$$

and in spherical basis:

$$F_{+1} = -\frac{1}{\sqrt{2}}(F_x + iF_y) = -\frac{r^2}{2\sqrt{3\pi}} I_1 \tag{4.242}$$

$$F_{-1} = -\frac{1}{\sqrt{2}}(F_x - iF_y) = -\frac{r^2}{2\sqrt{3\pi}} I_{-1} \tag{4.243}$$

$$F_0 = F_z = -\frac{r^2}{2\sqrt{3\pi}} I_0 \tag{4.244}$$

In the second stage, the integral in (4.236) should be solved. To this end, we have to evaluate this integral with $\vec{B}(\vec{r}, t)$ by considering (4.114):

$$\begin{aligned}
I_1 = \frac{1}{r^2} \int d\Omega Y_{11}(\vec{n}) \Bigg\{ \sum_{l,m} \frac{1}{c^{l+1}} \frac{\sqrt{4\pi(2l+1)(l+1)}}{\sqrt{l}(2l+1)!!} \\
\times \left[\sum_{n=0}^{\infty} \frac{1}{n!c^{2n}} M_{lm}^{(n)(l+2n+1)}\left(0, t - \frac{r}{c}\right) \right. \\
\times \left(\frac{\sqrt{l}}{\sqrt{2l+1}} \vec{Y}_{ll+1m}(\vec{n}) + \frac{\sqrt{l+1}}{\sqrt{2l+1}} \vec{Y}_{ll-1m}(\vec{n}) \right) \\
- iQ_{lm}^{(0)(l+1)}\left(0, t - \frac{r}{c}\right) \vec{Y}_{llm}(\vec{n}) + \frac{i}{c} \\
\left. \times \sum_{n=0}^{\infty} \frac{1}{n!c^{2n}} T_{lm}^{(n)(l+2n+2)}\left(0, t - \frac{r}{c}\right) \vec{Y}_{llm}(\vec{n}) \right] \Bigg\}
\end{aligned}$$

$$\times \left\{ \sum_{l'm'} \frac{1}{c^{l'+1}} \frac{\sqrt{4\pi(2l'+1)(l'+1)}}{\sqrt{l'}(2l'+1)!!} \right.$$
$$\times \left[\frac{1}{n'!c^{2n'}} M_{l'm'}^{(n')(l'+2n'+1)}\left(0, t-\frac{r}{c}\right) \right.$$
$$\times \left(\frac{\sqrt{l'}}{\sqrt{2l'+1}} \vec{Y}_{l'l'+1m'}(\vec{n}) + \frac{\sqrt{l'+1}}{\sqrt{2l'+1}} \vec{Y}_{l'l'-1m'}(\vec{n}) \right)$$
$$- Q_{l'm'}^{(0)(l'+1)}\left(0, t-\frac{r}{c}\right) \vec{Y}_{l'l'm'}(\vec{n})$$
$$\left. \left. + \frac{i}{c} \sum_{n'=0}^{\infty} \frac{1}{n'!c^{2n}} T_{l'm'}^{(n')(l'+2n'+2)}\left(0, t-\frac{r}{c}\right) \vec{Y}_{l'l'm'}(\vec{n}) \right] \right\} \tag{4.245}$$

Here, to calculate I_1, the integral should be split into different pieces and analyzed accordingly:

$$I_1 = I_1^{MM} + I_1^{QQ} + I_1^{TT} + I_1^{QM} + I_1^{TM} + I_1^{QT} \tag{4.246}$$

Since all pieces possess obvious corresponding forms, hence, we only provide the definition of I_1^{MM} below:

$$I_1^{MM} = \sum_{l,m,l',m'} \sum_{n,n'} \frac{1}{c^{l+l'+2n+2n'+2}} \frac{4\pi\sqrt{(l+1)(l'+1)}}{n!n'!\sqrt{ll'}(2l+1)!!(2l'+1)!!} M_{lm}^{(n)(l+2n+1)}$$
$$\times M_{l'm'}^{(n')(l'+2n'+1)} \times \left[\sqrt{ll'} \int d\Omega Y_{11} \vec{Y}_{ll+1m} \cdot \vec{Y}_{l'l'+1m'} + 2\sqrt{l(l'+1)} \right.$$
$$\left. \times \int d\Omega Y_{11} \vec{Y}_{ll+1m} \cdot \vec{Y}_{l'l'+1m'} + \sqrt{(l+1)(l'+1)} \int d\Omega Y_{11} \vec{Y}_{ll+1m} \cdot \vec{Y}_{l'l'+1m'} \right] \tag{4.247}$$

After this step, we may drop the arguments (0, t-r/c) and $\vec{n}$ from the spherical vector functions.

In the third step of the recoil force calculations, the number of surface integrals should be calculated over the products of three spherical harmonics, which can be performed in terms of $3j$ Wigner coefficients (Edmonds 1958), as:

$$\int d\Omega Y_{l_1m_1}(\vec{n}) Y_{l_2m_2}(\vec{n}) Y_{l_3m_3}(\vec{n}) = \frac{\sqrt{(2l_1+1)(2l_2+1)(2l_3+1)}}{2\sqrt{\pi}}$$
$$\times \begin{pmatrix} l_1 & l_2 & l_3 \\ 0 & 0 & 0 \end{pmatrix} \begin{pmatrix} l_1 & l_2 & l_3 \\ m_1 & m_2 & m_3 \end{pmatrix} \tag{4.248}$$

Using the following well-known relation:

$$\begin{pmatrix} l_1 & l_2 & l_3 \\ m_1 & m_2 & m_3 \end{pmatrix} = (-1)^{l_1-l_2-m_3} \frac{1}{\sqrt{2l_3+1}} C^{l_1,l_2,l_3}_{m_1,m_2,-m_3} \tag{4.249}$$

and expressing the Cartesian components of $\vec{Y}_{ll'm}(\vec{n})$ in terms of spherical harmonics $Y_{lm}(\vec{n})$ through 3j symbols, we can write:

$$\begin{aligned} I_1^{MM} &= \frac{1}{r^2} \sum_{l,n,n'} \frac{1}{c^{2l+2n+2n'+1}} \frac{8l(l+1)\sqrt{3\pi l}}{n!n'!(2l+1)!!(2l-1)!!\sqrt{(2l-1)(2l+1)}} \\ &\times \sum_m M_{lm}^{(n)(l+2n+1)} M_{l-1,-1-m}^{(n')(l+2n')} \begin{pmatrix} l & l-1 & 1 \\ m & -1-m & 1 \end{pmatrix} + \frac{1}{r^2} \\ &\times \sum_{l,m} \frac{(-1)^{l+1}}{c^{2l+2n+2n'+3}} \frac{4(l+2)(2l^2+4l+1)\sqrt{3\pi}}{n!n'!(2l+1)!!(2l+3)!!\sqrt{(l+1)(2l+1)(2l+3)}} \\ &\times \sum_m M_{lm}^{(n)(l+2n+1)} M_{l+1,-1-m}^{(n')(l+2n'+2)} \begin{pmatrix} l & l+1 & 1 \\ m & -1-m & 1 \end{pmatrix} \end{aligned} \tag{4.250}$$

$$\begin{aligned} I_1^{QQ} &= \frac{1}{r^2} \sum_{l,m} \frac{(-1)^l}{c^{2l+1}} \frac{2\sqrt{3\pi}(l+1)}{(2l-1)!!(2l-3)!!\sqrt{l(2l-1)(2l+1)}} \\ &\times \sum_m Q_{lm}^{(0)(l+1)} Q_{l-1,-1-m}^{(0)(l)} \begin{pmatrix} l & l-1 & 1 \\ m & -1-m & 1 \end{pmatrix} + \frac{1}{r^2} \\ &\times \sum_{l,m} \frac{(-1)^{l-1}}{c^{2l+3}} \frac{2\sqrt{3\pi}(l+2)}{(2l-1)!!(2l+1)!!\sqrt{(l+1)(2l+1)(2l+3)}} \\ &\times \sum_m Q_{lm}^{(0)(l+1)} Q_{l+1,-1-m}^{(n')(l+2)} \begin{pmatrix} l & l+1 & 1 \\ m & -1-m & 1 \end{pmatrix} \end{aligned} \tag{4.251}$$

$$\begin{aligned} I_1^{TT} &= \frac{1}{r^2} \sum_{l,n,n'} \frac{(-1)^l}{n!n'!c^{2l+2n+2n'+2}} \frac{2\sqrt{3\pi}(l+1)}{(2l-1)!!(2l-3)!!\sqrt{l(2l-1)(2l+1)}} \\ &\times \sum_m T_{lm}^{(n)(l+2n+2)} T_{l-1,-1-m}^{(n')(l+2n'+1)} \begin{pmatrix} l & l-1 & 1 \\ m & -1-m & 1 \end{pmatrix} + \frac{1}{r^2} \\ &\times \sum_{l,n,n'} \frac{(-1)^{l-1}}{n!n'!c^{2l+2n+2n'+4}} \frac{2\sqrt{3\pi}(l+2)}{(2l-1)!!(2l+1)!!\sqrt{(l+1)(2l+1)(2l+3)}} \\ &\times \sum_m T_{lm}^{(n)(l+2n+2)} T_{l+1,-1-m}^{(n')(l+2n'+3)} \begin{pmatrix} l & l+1 & 1 \\ m & -1-m & 1 \end{pmatrix} \end{aligned} \tag{4.252}$$

$$I_1^{QM} = \frac{4i}{r^2} \sum_{l,n} \frac{(-1)^{l-1}}{n!c^{2l+2n+2}} \frac{\sqrt{3\pi(l+1)(2l+1)}}{l\sqrt{l}(2l+1)!!(2l-1)!!}$$

$$\times \sum_m Q^{(n)(l+1)}_{l,-1-m} M^{(n)(l+2n+1)}_{l,m} \begin{pmatrix} l & l & 1 \\ -1-m & m & 1 \end{pmatrix} \tag{4.253}$$

$$I_1^{TM} = \frac{4i}{r^2} \sum_{l,n,n'} \frac{(-1)^l}{n!n'!c^{2n+2l+2n'+3}} \frac{\sqrt{3\pi (l+1)(2l+1)}}{l\sqrt{l}(2l+1)!!(2l-1)!!}$$

$$\times \sum_m T^{(n')(l+2n'+2)}_{l,-1-m} M^{(n)(l+2n+1)}_{l,m} \begin{pmatrix} l & l & 1 \\ -1-m & m & 1 \end{pmatrix} \tag{4.254}$$

$$I_1^{QT} = \frac{4}{r^2} \sum_{l,n'} \frac{(-1)^{l+1}}{n'!c^{2l+2n'+2}} \frac{\sqrt{3\pi}(l+1)}{(2l-1)!!(2l-3)!!\sqrt{l(2l+1)(2l-1)}}$$

$$\times \sum_m Q^{(0)(l+1)}_{lm} T^{(n')(l+2n'+1)}_{l-1,-1-m} \begin{pmatrix} l & l-1 & 1 \\ m & -1-m & 1 \end{pmatrix} + \frac{4}{r^2}$$

$$\times \sum_{l,n'} \frac{(-1)^l}{n'!c^{2l+2n'+4}} \frac{\sqrt{3\pi}(l+2)}{(2l-1)!!(2l+1)!!\sqrt{(l+1)(2l+1)(2l+3)}}$$

$$\times \sum_m Q^{(0)(l+1)}_{lm} T^{(n')(l+2n'+3)}_{l+1,-1-m} \begin{pmatrix} l & l+1 & 1 \\ m & -1-m & 1 \end{pmatrix} \tag{4.255}$$

So far, the calculation of the integral I_1 in (4.236) is presented.

Next, as the fourth step, we focus on the calculation of the integral I_{-1} in (4.237) through the use of the results that are obtained from (4.250) to (4.255), since:

$$I_{-1} = -I_1^* \tag{4.256}$$

Now, using the properties under complex conjugation of the double-superscript quantities $Q^{(n)(\upsilon)}_{lm}$ and $M^{(n)(\upsilon)}_{lm}$:

$$M^{(n)(\upsilon)*}_{lm} = (-1)^m M^{(n)(\upsilon)}_{l-m} \tag{4.257}$$

following from the same properties of the entire electric, magnetic, and toroidal form factors: $Q_{lm}(-k^2,t)$, $M_{lm}(-k^2,t)$, and $T_{lm}(-k^2,t)$. Changing m to $-m$ and following a similar strategy as we employed for I_1, we can consider the following relation (Edmonds 1958):

$$\begin{pmatrix} j_1 & j_2 & j_3 \\ m_1 & m_2 & m_3 \end{pmatrix} = (-1)^{j_1+j_2+j_3} \begin{pmatrix} j_1 & j_2 & j_3 \\ -m_1 & -m_2 & -m_3 \end{pmatrix} \tag{4.258}$$

In continue, as the fifth step, we define I_0 similar to I_1:

$$I_0 = I_0^{MM} + I_0^{QQ} + I_0^{TT} + I_0^{QM} + I_0^{TM} + I_0^{QT} \tag{4.259}$$

where each component can be described as:

$$I_0^{MM} = \frac{1}{r^2} \sum_{l,m,n,n'} \frac{1}{n!n'!c^{2l+2n+2n'+1}} \frac{8l(l+1)\sqrt{3\pi l}}{(2l-1)!!(2l+1)!!\sqrt{(2l-1)(2l+1)}}$$
$$\times M_{lm}^{(n)(l+2n+1)} M_{l-1,-m}^{(n')(l+2n')} \begin{pmatrix} l & l-1 & 1 \\ m & -m & 0 \end{pmatrix} + \frac{1}{r^2} \sum_{l,m} \frac{(-1)^{l+1}}{n!n'!c^{2l+2n+2n'+3}}$$
$$\times \frac{4\sqrt{3\pi}(l+2)(2l^2+4l+1)}{(2l+1)!!(2l+3)!!\sqrt{(l+1)(2l+1)(2l+3)}}$$
$$\times M_{lm}^{(n)(l+2n+1)} M_{l+1,-m}^{(n')(l+2n'+2)} \begin{pmatrix} l & l+1 & 1 \\ m & -m & 1 \end{pmatrix} \tag{4.260}$$

$$I_0^{QQ} = \frac{1}{r^2} \sum_{l,m} \frac{(-1)^l}{c^{2l+1}} \frac{2\sqrt{3\pi}(l+1)}{(2l-1)!!(2l-3)!!\sqrt{l(2l-1)(2l+1)}}$$
$$\times Q_{lm}^{(0)(l+1)} Q_{l-1,-m}^{(0)(l)} \begin{pmatrix} l & l-1 & 1 \\ m & -m & 0 \end{pmatrix} + \frac{1}{r^2} \sum_{l,m} \frac{(-1)^{l-1}}{c^{2l+3}}$$
$$\times \frac{2\sqrt{3\pi}(l+2)}{(2l-1)!!(2l+1)!!\sqrt{(l+1)(2l+1)(2l+3)}}$$
$$\times Q_{lm}^{(0)(l+1)} M_{l+1,-m}^{(0)(l+2)} \begin{pmatrix} l & l+1 & 1 \\ m & -m & 0 \end{pmatrix} \tag{4.261}$$

$$I_0^{TT} = \frac{1}{r^2} \sum_{l,m,n,n'} \frac{(-1)^l}{n!n'!c^{2l+2n+2n'+2}} \frac{2\sqrt{3\pi}(l+1)}{(2l-1)!!(2l-3)!!\sqrt{l(2l-1)(2l+1)}}$$
$$\times T_{lm}^{(n)(l+2n+2)} T_{l-1,-m}^{(n')(l+2n'+1)} \begin{pmatrix} l & l-1 & 1 \\ m & -m & 0 \end{pmatrix} + \frac{1}{r^2} \sum_{l,m} \frac{(-1)^{l-1}}{n!n'!c^{2l+2n+2n'+4}}$$
$$\times \frac{2\sqrt{3\pi}(l+2)}{(2l-1)!!(2l+1)!!\sqrt{(l+1)(2l+1)(2l+3)}}$$
$$\times T_{lm}^{(n)(l+2n+2)} M_{l+1,-m}^{(n')(l+2n'+3)} \begin{pmatrix} l & l+1 & 1 \\ m & -m & 1 \end{pmatrix} \tag{4.262}$$

$$I_0^{QM} = \frac{4i}{r^2} \sum_{l,m,n} \frac{(-1)^{l+1}}{n!c^{2l+2n+2}} \frac{\sqrt{3\pi(l+1)(2l+1)}}{(2l-1)!!(2l+1)!!l\sqrt{l}} Q_{l,-m}^{(0)(l+1)} M_{lm}^{(n)(l+2n+1)}$$
$$\times \begin{pmatrix} l & l & 1 \\ -m & m & 0 \end{pmatrix} \tag{4.263}$$

$$I_0^{TM} = \frac{4i}{r^2} \sum_{l,m,n,n'} \frac{(-1)^l}{n!n'!c^{2l+2n+2n'+3}} \frac{\sqrt{3\pi(l+1)(2l+1)}}{(2l-1)!!(2l+1)!!l\sqrt{l}} T_{l,-m}^{(n')(l+2n'+2)} M_{lm}^{(n)(l+2n+1)}$$
$$\times \begin{pmatrix} l & l & 1 \\ -m & m & 0 \end{pmatrix} \tag{4.264}$$

$$
\begin{aligned}
I_0^{QT} &= \frac{1}{r^2}\sum_{l,m,n'}\frac{(-1)^{l+1}}{n'!c^{2l+2n'+2}}\frac{4\sqrt{3\pi}(l+1)^2}{(2l-1)!!(2l-3)!!l\sqrt{l(2l-1)(2l+1)}}\\
&\times Q_{lm}^{(0)(l+1)}T_{l-1,-m}^{(n')(l+2n'+1)}\begin{pmatrix} l & l-1 & 1\\ m & -m & 0\end{pmatrix}+\frac{1}{r^2}\sum_{l,m,n'}\frac{(-1)^l}{n'!c^{2l+2n'+4}}\\
&\times\frac{4\sqrt{3\pi}(l+2)^2}{(2l-1)!!(2l+1)!!l\sqrt{(l+1)(2l+1)(2l+3)}}\\
&\times Q_{lm}^{(0)(l+1)}T_{l+1,-m}^{(n')(l+2n'+3)}\begin{pmatrix} l & l+1 & 1\\ m & -m & 0\end{pmatrix}
\end{aligned}\tag{4.265}
$$

Now, as a final step, we conclude the description of the recoil force for the general configuration of charges and currents. Using the spherical basis ($\mu = -1,0,1$) with the help of $3j$-symbols, the results can be expressed in terms of the derivatives of associated electric, magnetic, and toroidal form factors. Here, the order of the derivation with respect to momentum transfer ($-k^2$) is defined by the first superscript, whereas the order of time derivative is stated by the second superscript:

$$
F_\mu = [Q,Q]+[M,M]+[T,T]+[Q,M]+[Q,T]+[M,T]\tag{4.266}
$$

$$
\begin{aligned}
[Q,Q] &= \sum_l\frac{(-1)^{l+1}}{c^{2l+1}}\frac{(l+1)}{(2l-1)!!(2l-3)!!\sqrt{l(2l-1)(2l+1)}}\\
&\times\sum_m\begin{pmatrix} l & l-1 & 1\\ m & -\mu-m & \mu\end{pmatrix}Q_{lm}^{(0)(l+1)}Q_{l-1,-\mu-m}^{(0)(l+1)}\\
&+\sum_l\frac{(-1)^l}{c^{2l+3}}\frac{(l+2)}{(2l-1)!!(2l+1)!!\sqrt{(l+1)(2l+1)(2l+3)}}\\
&\times\sum_m\begin{pmatrix} l & l+1 & 1\\ m & -\mu-m & \mu\end{pmatrix}Q_{lm}^{(0)(l+1)}Q_{l+1,-\mu-m}^{(0)(l+2)}
\end{aligned}\tag{4.267}
$$

$$
\begin{aligned}
[M,M] &= \sum_{l,n,n'}\frac{(-1)^{l+1}}{n!n'!c^{2l+2n+2n'+1}}\frac{4l(l+1)\sqrt{l}}{(2l-1)!!(2l+1)!!\sqrt{(2l-1)(2l+1)}}\\
&\times\sum_m\begin{pmatrix} l & l-1 & 1\\ m & -\mu-m & \mu\end{pmatrix}M_{lm}^{(n)(l+2n+1)}M_{l-1,-\mu-m}^{(n')(l+2n')}\\
&+\sum_{l,n,n'}\frac{(-1)^l}{n!n'!c^{2l+2n+2n'+3}}\frac{2(l+2)(2l^2+4l+1)}{(2l+1)!!(2l+3)!!\sqrt{(l+1)(2l+1)(2l+3)}}\\
&\times\sum_m\begin{pmatrix} l & l+1 & 1\\ m & -\mu-m & \mu\end{pmatrix}M_{lm}^{(n)(l+2n+1)}M_{l+1,-\mu-m}^{(n')(l+2n'+2)}
\end{aligned}\tag{4.268}
$$

$$[T,T]=\sum_{l,n,n'}\frac{(-1)^{l+1}}{n!n'!c^{2l+2n+2n'+2}}\frac{(l+1)}{(2l-1)!!(2l-3)!!\sqrt{l(2l-1)(2l+1)}}$$
$$\times\sum_{m}\begin{pmatrix} l & l-1 & 1\\ m & -\mu-m & \mu\end{pmatrix}T_{lm}^{(n)(l+2n+2)}T_{l-1,-\mu-m}^{(n')(l+2n'+1)}$$
$$+\sum_{l,n,n'}\frac{(-1)^{l}}{n!n'!c^{2l+2n+2n'+4}}\frac{(l+2)}{(2l-1)!!(2l+1)!!\sqrt{(l+1)(2l+1)(2l+3)}}$$
$$\times\sum_{m}\begin{pmatrix} l & l+1 & 1\\ m & -\mu-m & \mu\end{pmatrix}T_{lm}^{(n)(l+2n+2)}T_{l+1,-\mu-m}^{(n')(l+2n'+3)} \tag{4.269}$$

$$[Q,M]=-2i\sum_{l,n}\frac{(-1)^{l}}{n!c^{2l+2n+2}}\frac{\sqrt{(l+1)(2l+1)}}{(2l-1)!!(2l-3)!!l\sqrt{l}}$$
$$\times\sum_{m}\begin{pmatrix} l & l & 1\\ -\mu-m & m & \mu\end{pmatrix}Q_{l,-\mu-m}^{(0)(l+1)}M_{lm}^{(n)(l+2n+1)} \tag{4.270}$$

$$[Q,T]=2\sum_{l,n'}\frac{(-1)^{l}}{n'!c^{2l+2n'+2}}\frac{(l+1)}{(2l-1)!!(2l-3)!!l\sqrt{l(2l-1)(2l+1)}}$$
$$\times\sum_{m}\begin{pmatrix} l & l-1 & 1\\ m & -\mu-m & \mu\end{pmatrix}Q_{lm}^{(0)(l+1)}T_{l-1,-\mu-m}^{(n')(l+2n'+1)}$$
$$+2\sum_{l,n'}\frac{(-1)^{l+1}}{n'!c^{2l+2n+2n'+4}}\frac{(l+2)}{(2l-1)!!(2l+1)!!\sqrt{(l+1)(2l+1)(2l+3)}}$$
$$\times\sum_{m}\begin{pmatrix} l & l+1 & 1\\ m & -\mu-m & \mu\end{pmatrix}Q_{lm}^{(0)(l+1)}T_{l+1,-\mu-m}^{(n')(l+2n'+3)} \tag{4.271}$$

$$[M,T]=-2i\sum_{l,n}\frac{(-1)^{l+1}}{n!n'!c^{2l+2n+2n'+3}}\frac{\sqrt{(l+1)(2l+1)}}{(2l-1)!!(2l+1)!!l\sqrt{l}}$$
$$\times\sum_{m}\begin{pmatrix} l & l & 1\\ -\mu-m & m & \mu\end{pmatrix}T_{l,-\mu-m}^{(n')(l+2n'+2)}M_{lm}^{(n)(l+2n+1)} \tag{4.272}$$

In the set of equations that have been employed in this Chapter, the following summations were utilized: $m=-l,\ldots,+l$; $l=1,2,\ldots$; $n,n=0,1,2,\ldots$. In addition, the emerging $3j$ Wigner coefficients are associated with the Clebsch-Gordan coefficients, given by Edmonds (1958):

$$\begin{pmatrix} j_1 & j_2 & j_3\\ m_1 & m_2 & m_3\end{pmatrix}=\frac{(-1)^{j_1-j_2-m_3}}{\sqrt{2j_3+1}}C_{m_1,m_2,-m_3}^{j_1,j_2,j_3} \tag{4.273}$$

Moreover, the Cartesian components of the recoil force F_x, F_y, F_z are correlated with $F_\mu(\mu=-1,0,1)$:

$$F_{+1} = -\frac{1}{\sqrt{2}}\left(F_x + iF_y\right) \tag{4.274}$$

$$F_{-1} = \frac{1}{\sqrt{2}}\left(F_x - iF_y\right) \tag{4.275}$$

$$F_0 = F_z \tag{4.276}$$

Ultimately, one could find the expression of the recoil force up to $1/c^5$ order from the general description of $F_\mu(\mu = -1,0,1)$:

$$F_\alpha = -\frac{2}{3c^4}\varepsilon_{\alpha\beta\gamma}\ddot{d}_\beta\dddot{m}_\gamma - \frac{1}{5c^5}\dddot{m}_{\alpha\beta}\dddot{m}_\beta - \frac{2}{5c^5}\overset{\therefore}{Q}_{\alpha\beta}\ddot{d}_\beta - \frac{2}{3c^5}\varepsilon_{\alpha\beta\gamma}\dddot{m}_\beta\overset{\therefore}{t}_\gamma \tag{4.277}$$

As can be seen in the equation above, besides the contributions from the classical electric and magnetic multipoles, the recoil force, started from the $1/c^5$ order, also possesses contributions from the toroidal dipole moment:

$$\vec{F}_{magnetic,toroidal} = -\frac{2}{3c^5}\overset{::}{\vec{m}} \times \overset{\therefore}{\vec{t}} \tag{4.278}$$

4.5 The Connection Between Cartesian and Spherical Components of the First Multipoles

In this section, we summarize the connections between the spherical and Cartesian components of the first multipoles. First, we present the electric ($\vec{d}$), magnetic ($\vec{m}$), and toroidal dipole ($\vec{t}$) moments, and electric ($Q_{\alpha\beta}$) and magnetic ($m_{\alpha\beta}$) quadrupole moments below.

For the electric multipole moment, we employed:

$$Q_{lm}(t) = \sqrt{\frac{4\pi}{2l+1}}\int d^3r r^l Y^*_{lm}\left(\frac{\vec{r}}{r}\right)\rho(\vec{r}, t) \tag{4.279}$$

Here, with $\vec{d} = \int \rho(\vec{r}, t)\vec{r}d^3r$, one can write: $Q_{10} = d_z$, $Q_{11} = 1\big/\sqrt{2}(-d_x + id_y)$, and $Q_{1-1} = 1\big/\sqrt{2}(d_x + id_y)$.

For the magnetic multipole moment, we considered:

$$M_{lm}(t) = -\frac{i}{c}\frac{\sqrt{4\pi l}}{\sqrt{(2l+1)(l+1)}}\int d^3r r^l \vec{Y}^*_{llm}\left(\frac{\vec{r}}{r}\right)\vec{j}(\vec{r}, t) \tag{4.280}$$

Here, with $\vec{m} = 1/2c\int[\vec{r}\times\vec{j}(\vec{r},t)]d^3r$, one can obtain:$M_{10} = -m_z$, $M_{11} = 1/\sqrt{2}(m_x - im_y)$, and $M_{1-1} = -1/\sqrt{2}(m_x + im_y)$.

For the toroidal multipole moment, we used:

$$T_{lm}(t) = -\frac{\sqrt{\pi l}}{c(2l+1)}\int d^3r r^{l+1}\left[Y^*_{ll-1m}\left(\frac{\vec{r}}{r}\right) + \frac{2}{(2l+3)}\times\frac{\sqrt{l}}{\sqrt{l+1}}Y^*_{ll+1m}\left(\frac{\vec{r}}{r}\right)\right]\vec{j}(\vec{r},t) \tag{4.281}$$

Here, with $\vec{t} = 1/10c\int[\vec{r}(\vec{r}\cdot\vec{j}) - 2\vec{r}^2\vec{j}]d^3r$, one can write:$T_{10} = t_z$, $T_{11} = -1/\sqrt{2}(t_x - it_y)$, and $T_{1-1} = 1/\sqrt{2}(t_x + it_y)$.

For the electric quadrupole moment with $l = 2$, we employed:

$$Q_{2m}(t) = \sqrt{\frac{4\pi}{5}}\int d^3r r^2 Y^*_{2m}\left(\frac{\vec{r}}{r}\right)\rho(\vec{r},t) \tag{4.282}$$

Here, with $Q_{ij} = 1/2\int\rho(\vec{r},t)[r_i r_j - 1/3\delta_{ij}\vec{r}]d^3r$, one can attain:$Q_{20} = 3Q_{zz}$, $Q_{22} = \sqrt{6}/2(Q_{xx} - 2iQ_{xy} - Q_{yy})$, $Q_{2-2} = \sqrt{6}/2(Q_{xx} + 2iQ_{xy} - Q_{yy})$, $Q_{2-1} = \sqrt{6}(Q_{xz} + iQ_{yz})$, and $Q_{2-2} = \sqrt{6}(-Q_{xz} + iQ_{yz})$.

For the magnetic quadrupole moment with $l = 2$, we considered:

$$M_{2m}(t) = -\frac{i}{c}\frac{\sqrt{8\pi}}{\sqrt{15}}\int d^3r r^2\vec{Y}^*_{22m}\left(\frac{\vec{r}}{r}\right)\vec{j}(\vec{r},t) \tag{4.283}$$

Here, with $\vec{m}_{ij} = 1/3c\int[(\vec{r}\times\vec{j})_i r_j + (\vec{r}\times\vec{j})_j r_i]d^3r$, one can obtain: $M_{20} = -3/2m_{zz}$, $M_{21} = \sqrt{3/2}(m_{xz} - im_{yz})$, $M_{2-1} = -\sqrt{3/2}(m_{xz} + im_{yz})$, $M_{22} = -\sqrt{6}/4(m_{xx} - 2im_{xy} - m_{yy})$, and $M_{2-2} = -\sqrt{6}/4(m_{xx} + 2im_{xy} - m_{yy})$.

For the toroidal quadrupole moment with $l = 2$, we took into account:

$$T_{2m}(t) = -\frac{\sqrt{2\pi}}{5c}\int d^3r r^3\left[Y^*_{21m}\left(\frac{\vec{r}}{r}\right) + \times\frac{2\sqrt{2}}{7\sqrt{3}}Y^*_{23m}\left(\frac{\vec{r}}{r}\right)\right]\vec{j}(\vec{r},t) \tag{4.284}$$

Here, with $\vec{t}_{ik} = 1/28c\int d^3r[4r_i r_k(\vec{r}\cdot\vec{j}) - 5r^2(r_i j_k + r_k j_i) + 2r^2(\vec{r}\cdot\vec{j})\delta_{ik}]$, one can write:$T_{20} = t_{zz}$, $T_{21} = -\sqrt{2}/\sqrt{3}(t_{xz} - it_{yz})$, $T_{2-1} - \sqrt{2}/\sqrt{3}(t_{xz} + it_{yz})$, $T_{22} = 1/\sqrt{6}(t_{xx} - 2it_{xy} - t_{yy})$, and $T_{2-1} = 1/\sqrt{6}(t_{xx} + 2it_{xy} - t_{yy})$.

For the electric octupole moment with $l = 3$, we considered:

$$Q_{3m}(t) = \sqrt{\frac{4\pi}{7}} \int d^3 r r^3 Y_{3m}^* \left(\frac{\vec{r}}{r}\right) \rho(\vec{r}, t) \tag{4.285}$$

Here, with $Q_{ijk} = 1/6 \int \rho(\vec{r}, t)[r_i r_k r_j - 1/5 r^2 (r_i \delta_{kl} + r_k \delta_{il} + r_l \delta_{ik})] d^3 r$, one can attain: $Q_{30} = 15 Q_{zzz}, Q_{31} = -15\sqrt{3}/2(Q_{zzx} + i Q_{yyy} + i Q_{xxy}), Q_{32} = -3\sqrt{15/2}(Q_{zzz} + 2Q_{yyz} + 2i Q_{xyz})$, $Q_{32} = -3\sqrt{5/2}(Q_{xxx} - 3Q_{yyx} + i Q_{xyz} - 3i Q_{xxy})$, and $Q_{3,-m} = (-1)^m Q_{3m}^*$.

For the magnetic octupole moment with $l = 3$, we employed:

$$M_{3m}(t) = -\frac{i}{c} \sqrt{\frac{3\pi}{7}} \int d^3 r r^2 \vec{Y}_{33m}^* \left(\frac{\vec{r}}{r}\right) \vec{j}(\vec{r}, t) \tag{4.286}$$

Here, with $\vec{m}_{ijk} = 15/2c \int d^3 r [r_i r_j (\vec{r} \times \vec{j})_k + r_i r_k (\vec{r} \times \vec{j})_j + r_j r_k (\vec{r} \times \vec{j})_i + \delta_{ij}/5(\vec{r} \times \vec{r} \times \vec{r} \times \vec{j})_k + \delta_{ik}/5(\vec{r} \times \vec{r} \times \vec{r} \times \vec{j})_j$, one can obtain: $M_{30} = -1/22 m_{zzz}$, $M_{31} = 1/8\sqrt{3}(m_{zzx} + i m_{yyy} + i m_{xxy})$, $M_{32} = \sqrt{2}/8\sqrt{15}(m_{zzz} + 2m_{yyz} + 2i m_{xyz})$, $M_{33} = 1/24\sqrt{5}(m_{xxx} - 3m_{yyx} + i m_{yyy} - 3i m_{xxy})$, and $M_{3-m} = (-1)^m M_{3m}^*$.

Moreover, the first mean-square radius of the electric for $l = 0$ is:

$$m = 0, \bar{r}_{00}^2(t) = \int d^3 r r^2 \rho(\vec{r}, t) \tag{4.287}$$

Here, with the Cartesian definition of $\bar{r}_q^{(2)}(t) = \int d^3 r r^2 \rho(\vec{r}, t)$, one can write $\bar{r}_q^{(2)}(t) = \bar{r}_{00}^2(t)$. The same component of the electric dipole for $l = 1$ is:

$$\bar{r}_{1m}^2(t) = \sqrt{\frac{4\pi}{3}} \int d^3 r r^3 \rho(\vec{r}, t) Y_{1m}^* \left(\frac{\vec{r}}{r}\right) \tag{4.288}$$

Similarly, with the Cartesian definition of $\vec{\bar{r}}_d^{(2)}(t) = \int d^3 r \vec{r} r^2 \rho(\vec{r}, t)$, one can acquire $\bar{r}_{10}^2 = (\bar{r}_d^2) z, \bar{r}_{1-1}^2 = 1/\sqrt{2}[(\bar{r}_d^2)_x + i(\bar{r}_d^2)_y]$.

On the other hand, the first mean-square radius of the magnetic dipole distribution for $l = 1$ is:

$$\bar{\rho}_{1m}^2 = \frac{1}{ic} \sqrt{\frac{2\pi}{3}} \int d^3 r r^3 \vec{j}(\vec{r}, t) Y_{1m}^* \left(\frac{\vec{r}}{r}\right) \tag{4.289}$$

Here, with the Cartesian definition of $\vec{\bar{\rho}}^2 = 1/2c \int d^3 r r^2 (\vec{r} \times \vec{j})$, one can obtain $\bar{\rho}_{10}^2 = -\bar{\rho}_z^2, \bar{\rho}_{11}^2 = 1/\sqrt{2}(\bar{\rho}_x^2 - i\bar{\rho}_y^2)$, and $\bar{\rho}_{1-1}^2 = 1/\sqrt{2}(-\bar{\rho}_x^2 - i\bar{\rho}_y^2)$.

Finally, the first mean-square radius of the toroidal dipole for $l = 1$ can be defined as:

$$\bar{R}_{1m}^2(t) = -\frac{\sqrt{2\pi}}{3c}\int d^3 r r^4 \vec{j}(\vec{r}, t)\left[\frac{1}{7}Y_{12m}^*\left(\frac{\vec{r}}{r}\right) + \frac{1}{2\sqrt{2}}Y_{10m}^*\left(\frac{\vec{r}}{r}\right)\right] \quad (4.290)$$

and in the Cartesian basis, this results in the first mean-square radius of the toroidal dipole distribution as $\vec{\bar{R}}^2(t) = 1/28c\int d^3 r r^2\{3r^2\vec{j}(\vec{r}, t) - 2\vec{r}[\vec{r}\cdot\vec{j}(\vec{r}, t)]\}$, where one can write $\bar{R}_z^2 = -\bar{R}_{10}^2$, $\bar{R}_x^2 = 1/\sqrt{2}(\bar{R}_{11}^2 - \bar{R}_{1-1}^2)$, and $\bar{R}_y^2 = i/\sqrt{2}(\bar{R}_{11}^2 - \bar{R}_{1-1}^2)$.

References

V.M. Dubovik, A.A. Cheshkov, Multipole expansion in classic and quantum field theory and radiation. Sov. J. Particles. Nucl. **5**, 318–337 (1974)

V.M. Dubovik, V.V. Tugushev, Toroid moments in electrodynamics and solid-state physics. Phys. Rep. **187**, 145–202 (1990)

A. Edmonds, *Deformations of Atomic Nuclei* (Publishing House for Foreign Literature, Moscow, 1958)

C.G. Gray, Multipole expansions of electromagnetic fields using Debye potentials. Am. J. Phys. **46**(2), 169–179 (1978)

L.D. Landau, E.M. Lifshitz, *Course of Theoretical Physics*, vol. 2 (Pergamon Press, New York, 1993)

E.E. Radescu, G. Vaman, Exact calculation of the angular momentum loss, recoil force, and radiation intensity for an arbitrary source in terms of electric, magnetic, and toroid multipoles. Phys. Rev. E **65**, 046609 (2002)

Chapter 5
Toroidal Excitations in Metamaterials

Abstract Thus far, we showed that toroidal excitations exist in free space as spatially and temporally confined electromagnetic pulses propagating at the velocity of light and interacting with matter. In this Chapter, we presented an exhaustive study on the theoretical and experimental observation of toroidal excitations in both bulk and quasi-infinite artificially structured media, also known as *metamaterials*. Using the established framework to analyze the toroidal electrodynamics, we discussed the strategies that have been utilized to efficiently excite toroidal modes in well-engineered subwavelength architectures. We initially argued the formation of the toroidal resonances in 3D metamaterials, and later, we revealed that how the flatland metaphotonics successfully addressed the fabrication, simulation, and inherent losses in bulk metastructures.

5.1 Toroidal Excitations in 3D Artificial Media

Metamaterials are artificially engineered periodic arrays of nano- and microscale building blocks that offer intriguing properties beyond natural materials. Applications of them, including but not limited to semiconductor and high-temperature superconductor characterization, asymmetric transmission, optical magnetism, hyperbolic dispersion, epsilon near-zero (ENZ), topological states, arbitrary control of light's trajectories, transformation optics, novel quantum information, tomographic imaging, label-free genetic analysis, cellular level imaging and chemical and biological sensing, have thrust metamaterials research from a relative obscurity into the limelight (Pendry et al. 2006; Boltasseva and Atwater 2011; Kildishev et al. 2013; Zheludev and Kivshar 2012; Soukoulis and Wegener 2011; Gerislioglu and Ahmadivand 2020; Gerislioglu et al. 2020).

In particular, to date, several types of metastructures have been designed and introduced in the field of plasmonics and photonics, to affectively manipulate light-matter interactions and support pronounced resonant states. Among them, electromagnetically induced transparency (EIT)- and Fano-resonant metamaterials have received copious interest in the first two-decades of the present Century, because of their

A. Ahmadivand et al., *Toroidal Metamaterials*, Engineering Materials,
https://doi.org/10.1007/978-3-030-58288-3_5

narrow spectral linewidth and the ability of controlling the direction of the transmitted/reflected light (e.g., beam steering) (Luk'yanchuk et al. 2010; Galarreta et al. 2018; Papasimakis et al. 2008; Gerislioglu et al. 2017). Metamaterials have proven their promise towards developing effective tunable optics-based instruments, but the need for attaining the most efficient and reliable design have urged the researchers to explore novel alternatives.

On the other hand, the topic of toroidal excitations with unparalleled properties was transferred from atomic, nuclear, molecular, and condensed matter physics studies to the classical optics through the use of metamaterials (Marinov et al. 2007). As mentioned in Chap. 2, toroidal multipoles are unconventional electromagnetic excitations, which are different from the ones related to traditional electric and magnetic multipoles. Theoretically, toroidal moments are missing from the standard multipole expansion (Silenko 1999; Dubovik and Tugushev 1990), and specifically, the dynamic toroidal dipole (i.e., the hidden counterpart of electric and magnetic dipoles) is originated from the poloidal currents flowing on the surface of a torus along its meridians, or it is equivalently considered as a head-to-tail loop of magnetic dipoles (Silenko 1999; Kaelberer et al. 2010). Toroidal dipole is the dominant member of the toroidal multipoles family and recognized for revealing the spatial feature of toroidal dipole mode as gyrotropic-fashioned charge-current excitations, in which their far-field radiation patterns are considerably concealed (Papasimakis et al. 2009; Ahmadivand et al. 2017; Fedotov et al. 2013).

In 2010, the initial experimental verification of the dynamic toroidal dipole mode is reported in a 3D metamaterial in the range of microwave frequencies (Kaelberer et al. 2010). The corresponding 3D design is illustrated in Fig. 5.1a, where a set of four rectangular, electrically isolated metallic wire loops are embedded within a low-loss dielectric layer. Here, to suppress the induced electric and magnetic moments and to enhance the toroidal dipolar response, this particular metamolecule layout is tailored. The corresponding geometric parameters of the metamolecule are $a = 1.8$ mm, $h = 1.5$ mm, $r = 2.44$ mm, $w = 0.15$ mm, $d = 8$ mm, $s = 7.5$ mm, and $g = 0.15$ mm. The assembled metamaterial slab was fabricated through high-resolution printed board technology with the overall dimension of 8 mm $\times$ 176 mm $\times$ 165 mm (Fig. 5.1b). The rows of the array were constructed from metalized microwave laminate strips and judiciously stacked at a regular interval with the axes of the metamolecules aligned along the plane of the metamolecules array. It is important to underline that before conducting the measurements, the green solder resist was removed from the samples. As depicted in Fig. 5.1a, the front and rear pairs of the metallic loops interact with the magnetic component of the incident light, which creates circular currents based on Faraday's law of induction. Indeed, the incoming wave travels through the front and rear metallic loops with a phase delay, and this can be understood from in-phase (parallel) and out-of-phase (antiparallel) components of the magnetic field at the loops. The unique result of this process is the formation of toroidal and magnetic resonances in the transmission (and reflection) spectra, which are labeled as '**II**' and '**I**', respectively (see Figs. 5.1c, d). At '**I**', the metamolecule interacts with the in-phase part of the magnetic field and the individually-formed magnetic moments (**m**) indicate the same direction along the

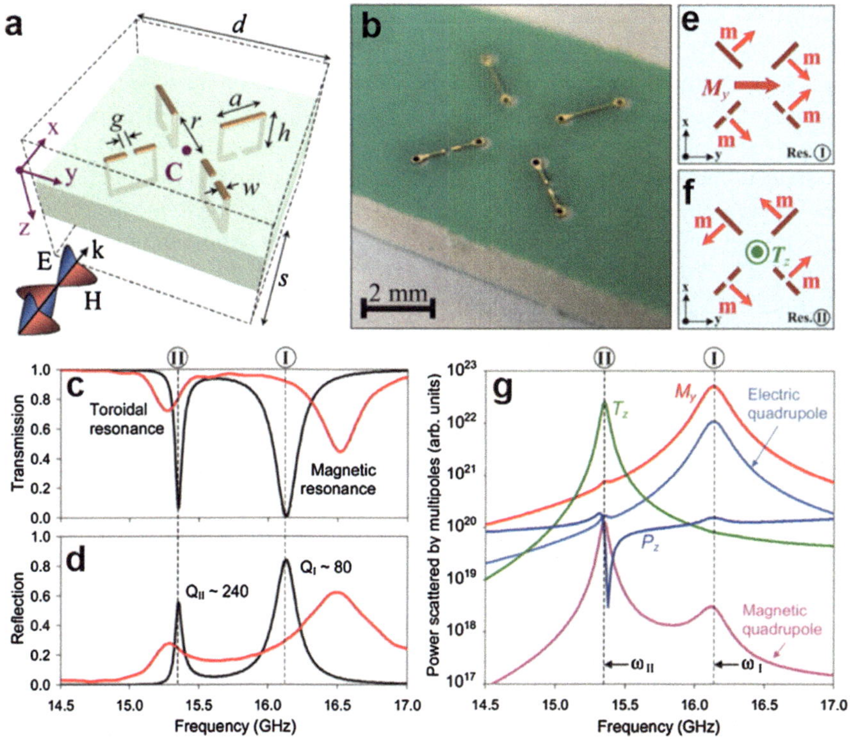

Fig. 5.1 Design and spectral response of the 3D toroidal metamolecule. **a** Schematic of the unit cell of the metamaterial, consisting of four split wire loops with respect to the incident wave. **b** An image of the assembled metamaterial slab. Experimentally measured (red line) and numerically calculated (black line) **c** transmission and **d** reflection spectra of the metamaterial. **e** and **f** Optically-driven magnetic moments (**m**) of wire loops during in-phase and out-of-phase interactions, respectively. **g** Multipolar decomposition spectra of the radiated power from different multipoles. Adapted from (Kaelberer et al. 2010) with permission. Copyright AAAS

y-axis, which contribute to magnetic dipole mode parallel to the incident magnetic field (Fig. 5.1e). Conversely, at '**II**', the metamolecule interacts with the out-of-phase part of the magnetic field and the individually-induced magnetic moments (**m**) form the head-to-tail loop configuration (Fig. 5.1f), which induce the toroidal dipole mode along the *z*-axis (T_z).

As discussed in Chap. 3, these qualitative claims and the nature of the induced resonances can be quantitatively assessed through the multipole decomposition analysis (Fig. 5 g). As can be seen from the scattered power plot, at '**II**' (~15.4 GHz), the resonance scattering is dominated by T_z. Obviously, the magnetic dipole and electric quadrupole excitations of the metamolecule are not resonant at the frequency of toroidal dipole. Therefore, one can claim that the typical multipole excitations cannot be responsible for the resonance feature at 15.4 GHz. This indicates that the resonance can be originated from a toroidal dipole, which scatters substantially compared to any of the conventional multipoles by almost two orders of magnitude. On the other hand, at 'I' (~16.1 GHz), while the other classical and unconventional multipoles are suppressed, the magnetic dipole (M_y) is the dominant component.

Primary investigations on the toroidal dipole rapidly captured attention that led to the development of diverse types of resonant architectures along a broad range of frequencies. Among them, in the 3D metamolecules limit, several toroidal micro and nanostructures have been designed to operate from the microwave frequencies to the near-infrared (NIR) wavelengths (Papasimakis et al. 2009; Ahmadivand et al. 2017; Fedotov et al. 2013; Fan et al. 2013; Liu et al. 2017; Yang et al. 2017). Figure 5.2 presents a 3D toroidal metamaterial based on an assembly of dumbbell-shaped aperture on a metallic surface, with fourfold and eightfold symmetries. A photograph of the assembled unit cell with eightfold symmetry is demonstrated in Fig. 5.2a. Figure 5.2b represents a metal screen with a dumbbell-shaped aperture, which is the fundamental element of toroidal metamaterial. This graph indicates the polarization of the incident beam and the axis of the unit cell's mirror symmetry.

In such a complex structure, the incident electromagnetic wave induces high-order multipoles, most notably magnetic quadrupole moment, which stems from the creation of a pair of counterrotating surface current densities. Figures 5.2c, d exhibit how the scattering contribution from this multipolar mode can be suppressed in the unit cell of fourfold or eightfold symmetries . This was shown analytically through

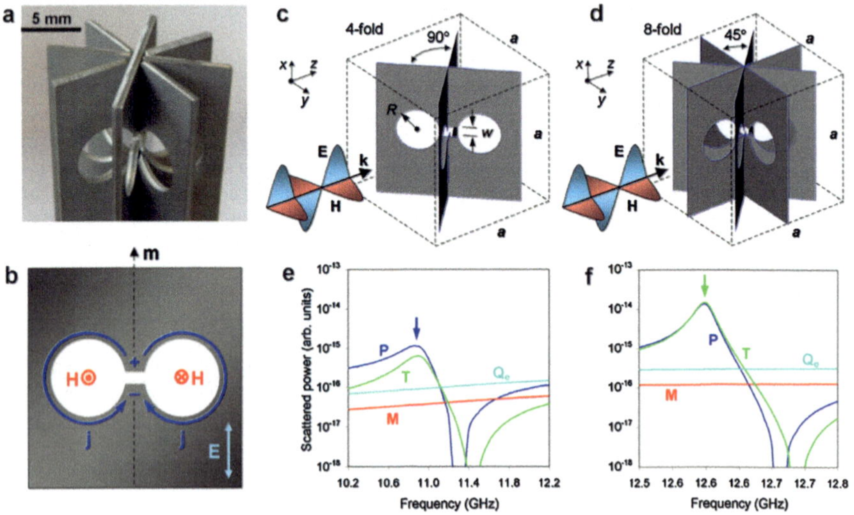

Fig. 5.2 Design and spectral response of the 3D toroidal metamolecule. **a** Photograph of one of the array's column with eightfold symmetry. **b** Metallic screen with a dumbbell-shaped surface. Metamaterial's unit cell with **c** fourfold and **d** eightfold symmetry. Dispersions of multipolar radiation rates calculated for 4 strongest multipoles induced in **e** fourfold and **d** eightfold symmetric metamolecules. Arrows indicate locations of the corresponding transparency resonances. Adapted from (Fedotov et al. 2013) with permission. Copyright NPG

modeling the interaction of the metamolecules configured in 2D arrays based on slabs of toroidal metamaterials with normally incident linearly polarized light. The computed densities of the induced currents can be employed to estimate the radiated power of the multipoles corresponding to each unit cell in the metamaterial. For instance, in the metamaterial with fourfold symmetric unit cells, calculations showed that except the electric dipole, the scattering of all other classical multipoles is negligible (Fig. 5.2e). In this set of analyses, the toroidal dipole is the dominant contributor at the resonance frequency. The same trend was also observed for the emitted power from multipoles of much more complex eightfold symmetric metamolecule (Fig. 5.2f).

Until now, we summarized some of the important achievements in the past decade for the development of toroidal metamaterials at the microwave frequencies. Recent advancements in both numerical analysis and fabrication of complex subwavelength structures have led to the emergence of novel 3D toroidal metamaterials at shorter wavelengths or higher frequencies. One of the recent examples is the toroidal metamaterial composed of two conductively joint gold split ring resonators on the edge of a square self-supporting gold membrane (Fig. 5.3a). Figure 5.3b demonstrates the scanning electron microscope (SEM) image of the fabricated toroidal metamaterial. This structure is tailored in such a way that to sustain double toroidal moments in the mid-infrared wavelengths. To motivate the design, a *y*-polarized beam was used to induce the plasmon resonances at the metal-air interface. The transmission spectra for both simulations and measurements in Figs. 5.3c, d, respectively, show the excitation of three pronounced minima correlating with two toroidal modes at $\omega_{T1} = 73$ THz and $\omega_{T2} = 83.5$ THz, and an electric dipole mode at $\omega_D = 98$ THz.

To quantitatively investigate the resonant properties of the plasmonic metamaterial, the radiated power of multipole moments was calculated from the induced volume current density in the metamaterial, displayed in Fig. 5.3e. Obviously, the electric/magnetic dipole, electric/magnetic quadrupole, and toroidal dipole are the major and dominant contributors to the metamaterial's response. Although both electric and magnetic quadrupole moments are resonant at the two resonance frequencies, the intensity of the toroidal component (T_z) at ω_{T1} and ω_{T2} is much stronger than that of classical multipoles. Importantly, one should note that the dominant toroidal component does not contribute directly to the far-field radiation of the transmission spectra at normal incidence, however, indirectly contributes through the mediation of the surface plasmon mode of the split resonators. Plotting the magnetic field and current distribution at the resonance frequencies allows to understand the generation mechanism of the resonant modes (Figs. 5.3f, h). For example, at ω_{T1}, one can observe the formation of a strong and tight magnetic vortex in the *x*-plane. Figures 5.3g, i demonstrate the configuration of the surface current distribution within the unit cell, which imply that the loop currents and accordingly the magnetic dipoles in the left part of the metamolecule and right part of the proximal one are opposite to each other. Such a head-to-tail configuration of the magnetic dipoles leads to a substantial coupling between the adjacent resonators, where the circulating magnetic field creates toroidal dipolar excitation oriented along the *z*-axis. On the other hand, at ω_{T2}, the calculated magnetic field was generated magnetic vortices that are firmly

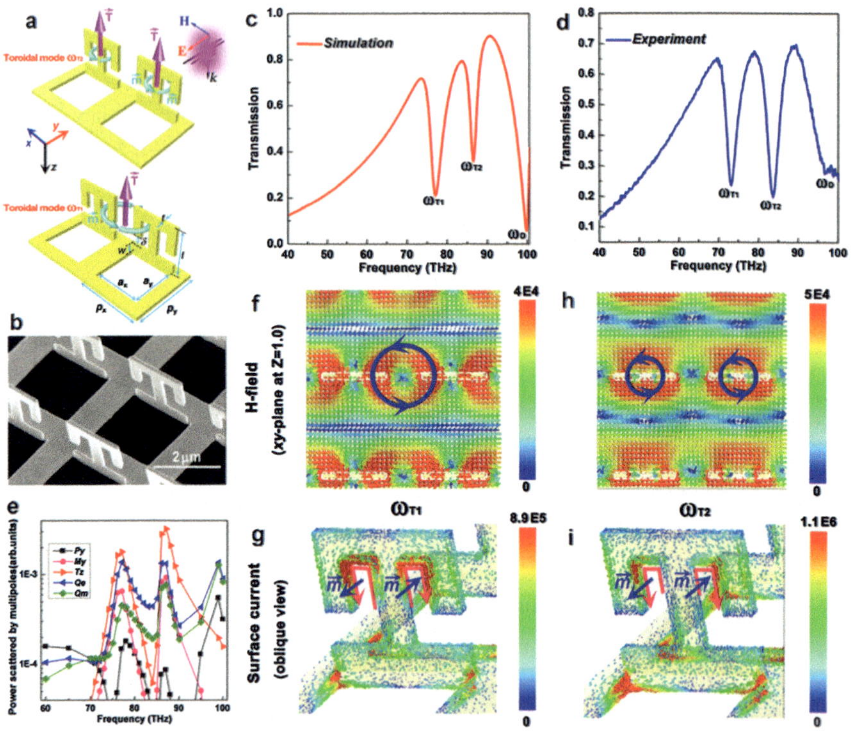

Fig. 5.3 Design and spectral response of the 3D toroidal metamolecule. **a** Schematic illustration of the 3D toroidal metamolecule with the following geometries: $a_x = a_y = 2\ \mu\text{m}$, $l = 1.98\ \mu\text{m}$, $w = 400$ nm, and $\delta = 500$ nm; periodicity is 3 μm and thickness of the gold film is $t = 150$ nm. **b** SEM image of the fabricated 3D toroidal metamaterial. **c** The simulated and **d** measured transmission spectra of the plasmonic metamaterial for a y-polarized beam illumination. **e** Dispersion of scattered power for several multipole moments induced in the metamaterial. **f** and **h** The numerically obtained magnetic field distribution of four unit cells of the toroidal metamaterial at ω_{T1} and ω_{T2}, respectively. **g** and **i** Surface current distribution and magnetic dipole orientation on a unit cell of the metamaterial at ω_{T1} and ω_{T2}, respectively. Adapted from (Yang et al. 2017) with permission. Copyright ACS

confined within the unit cell. The induced currents also form circulating surface currents along each split ring resonator, result in a magnetic dipole which points in forward or backward direction perpendicular to the joint double split ring resonators' plane. In this limit, opposite circular currents in the left and right side of each split ring resonators generate a circular magnetic moment distribution that is perpendicular to the structure's surface, possesses a strong toroidal moment towards the z-direction.

In recent years, several types of all-dielectric and plasmonic 3D metaplatforms have been devised and fabricated to support pronounced toroidal dipoles across a broad range of frequencies. However, limited access to the stored energy, volumetric losses, challenging lithography procedures, complex numerical and theoretical analysis requirements, and incompatible integration with planar devices have encouraged researchers to develop toroidal metastructures based on flatland metaphotonics.

5.2 Toroidal Multipoles in Planar Artificial Media

In the flatland optics regime, quasi-infinite metasurfaces, made of periodic arrays of subwavelength scatterers or optical thin films, surmount the need for propagation effect by enforcing abrupt and controllable changes of the optical properties. Straightforward and cost-effective fabrication steps, simple modeling and quick numerical computations, compatibility with planar devices in photonic/plasmonic circuits, as well as reduced radiative losses have inspired researchers to develop artificial media based on flatland optics. Besides conventional resonances, toroidal excitations have also been successfully realized using planar metamaterials. The excitation mechanism of a toroidal dipole in quasi-infinite structures can be understood by considering the behavior of the excited classical modes in the system. Generally, a magnetic dipole weakly couples to the free space relative to the electric dipole (Jing et al. 2018; Zhang et al. 2019). The formation of a toroidal dipole is the direct result of putting these magnetic dipoles into head-to-tail configurations. This reduces the coupling efficiency of the metasurface to the free space, which narrows the linewidth of the induced resonances and subsequently reduces the radiative losses of the entire system. Following parts argues the adopted schemes to excite toroidal dipole mode in plasmonic and all-dielectric metasurfaces, which encompasses both long and short wavelengths.

In continue, we consider two types of planar structures in symmetric and antisymmetric orientations. To begin with, we consider a planar toroidal metasurface consisting of periodic arrays of antisymmetric plasmonic meta-atoms (Figs. 5.4a, b) (Ahmadivand and Gerislioglu 2018). Modelling of the unit cell was performed by employing the empirically defined aluminum permittivities by assuming the presence of a thin ($\sim$2 nm) alumina (Al_2O_3) layer as a coverage coating of the resonators. Similar to the 3D metamaterials, multipole expansion analysis is required to extract the spectral response of planar metastructures. Figure 5.4c evaluates the contribution of the strongest multipoles to far field, indicating the dominant behavior of the toroidal dipole mode at 975 and 1450 nm, where all other classical multipoles are suppressed. Plotting the transmission and reflection spectra explicitly verifies the excitation of pronounced toroidal dipole resonances at the mentioned wavelengths (Fig. 5.4d). Figures 5.4e, f illustrate the electric-field (E-field) maps for the excitation of toroidal dipoles in a logarithmic scale. The ultratight confinement of the excited fields at the capacitive openings between the central and proximal resonators is an explicit result of the excitation of toroidal dipoles. Interestingly, at 975 nm, the field confinement is at the central part of the unit cell, and conversely, at 1450 nm, the fields are localized at the tips of the outermost of the unit cell. Figure 5.4g illustrates the artistic illustrations for the rotation of poloidal currents and the direction of the toroidal field, squeezed between the neighboring and middle resonators.

To further analyze the excitation principle of the toroidal modes, one needs to calculate the surface current density of the unit cell. The vectorial surface current density maps at the toroidal resonance wavelengths are depicted in Fig. 5.4h, showing a destructive mismatch between the induced magnetic moments in each resonator

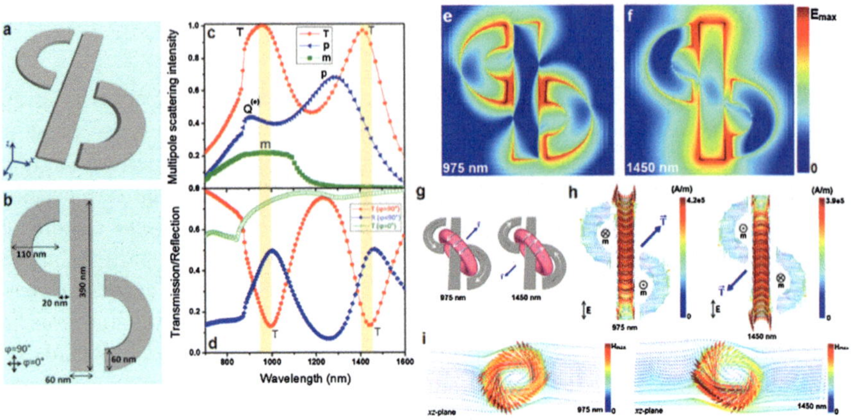

Fig. 5.4 Design and spectral response of the directional planar toroidal meta-atom. **a** Schematic representation of the planar plasmonic meta-atom. **b** Judiciously defined geometrical sizes of the meta-atom. **c** Multipole radiation intensity of the strongest moments. **d** Normalized transmission and reflection spectra emitted from the metasurface. **e** and **f** E-field maps of the toroidal dipole resonances in the meta-atom in logarithmic scale. **g** Representations of the formation of charge-current arrangements and poloidal currents within the structure. **h** Vectorial surface current density maps for the directional toroidal modes. **i** Cross-sectional (*xz*-plane) vectorial H-field maps to verify the formation of a head-to-tail configuration at the corresponding toroidal resonance wavelengths. Adapted from (Ahmadivand et al. 2018) with permission. Copyright ACS

required for the formation of toroidal modes. This effect uncovers the formation of oppositely pointed toroidal dipoles at different wavelengths, which can be verified by plotting the cross-sectional magnetic field (H-field) maps (Fig. 5.4i).

The excitation of pronounced and strong toroidal dipoles is also possible in unit cells with in-plane symmetry. Plasmonic oligomer nanocavities (Ögüt et al. 2012) and multipixel resonators (Gerislioglu et al. 2018; Ahmadivand et al. 2018) are some of the noteworthy examples of this principle. In Fig. 5.5, the spectral properties of a symmetric planar toroidal metamolecule is presented, consisting of two metallic pixels to support a strong toroidal dipole at the THz frequencies. An artistic rendering of the unit cell and its important geometrical parameters ($R_i/W/L/D/g$ = 60/15/105/15/5 μm) are indicated in Fig. 5.5a. The SEM image of the metasurface is depicted in Fig. 5.5b. The top-view E-field map of the plasmonic unit cell at the toroidal dipole frequency is shown in Fig. 5.5c, exhibiting the strong E-field intensity and localization at the capacitive openings of the unit cell. Formation of the required oppositely aligned magnetic moments within the adjacent resonators of the metamolecule is displayed in Fig. 5.5d. This panel clearly shows the antiparallel current distribution due to incident THz beam, which has undeniable role in inducing strong toroidal dipole. Figure 5.5e illustrates the cross-sectional vectorial H-field intensity profile, which confirms the closed-loop magnetic field around the central arms of the unit cell. Lastly, numerically predicted and experimentally measured transmission

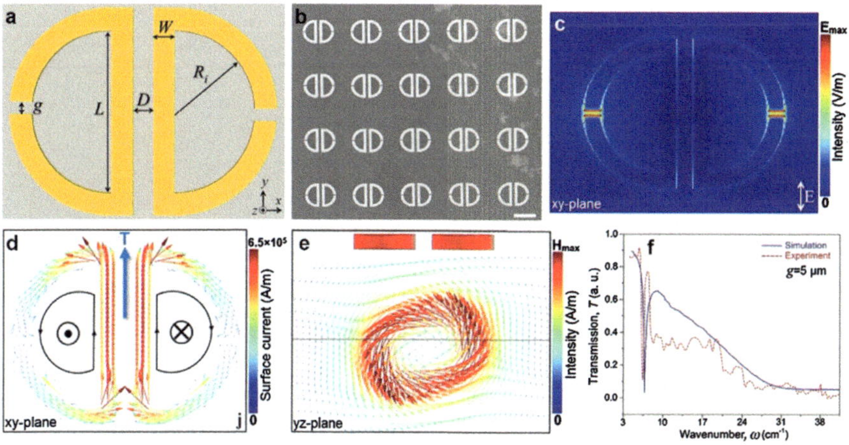

Fig. 5.5 Design and spectral properties of a plasmonic THz metasurface. **a** Top-view image and geometrical parameters of the unit cell. **b** SEM image of the metasurface: scale bar is 100 μm. **c** Local near-field map of the E-field enhancement at the gaps. **d** Surface current plot across the metastructure and formation of oppositely rotating magnetic fields. **e** *yz*-plane of the resonators, showing the formation of a head-to-tail magnetic field. **f** Numerically obtained and experimentally measured normalized transmission amplitude for the toroidal metasurface. Adapted from (Ahmadivand et al. 2018) with permission. Copyright OSA

spectra are plotted in Fig. 5.5f, showing the excitation of a profoundly sharp toroidal dipole around $\omega \sim 6\ \text{cm}^{-1}$.

While metallic structures at subwavelength dimensions have always been a promising choice for the development of metamaterials and metasurfaces, in recent years, all-dielectric metastructures have been acknowledged as alternative candidates to tailor efficient and cost-effective platforms. Such an interest stems from the low-loss characteristics of high-index resonators, where in an electric metal, the real part of the dielectric permittivity at optical frequencies is negative ($\varepsilon' < 0$). Similarly, in a magnetic metal, the real component of the magnetic permeability is negative ($\mu' < 0$). As can be seen, metallic structures are crucial components for obtaining either negative permeability or permittivity media. However, the internal damping and high-energy dissipation have led researchers to address these drawbacks through the use of all-dielectric subwavelength structures based on high-index materials (e.g., silicon, germanium, tellurium, and titanium dioxide). In principle, the interaction of light with frequency below or near the bandgap of the high-index material results in the excitation of both electric and magnetic resonances in the system. Recent studies have successfully reported the excitation of strong toroidal moments in high-index all-dielectric architectures (Basharin et al. 2015; Tuz et al. 2018; Sayanskiy et al. 2018; Xu et al. 2019).

Utilizing the exquisite properties of all-dielectric nanodisks, it is shown that a four-member quadrumer cluster can be utilized to sustain toroidal dipolar mode at the NIR wavelengths (Tuz et al. 2018). Figure 5.6a demonstrates a schematic of the

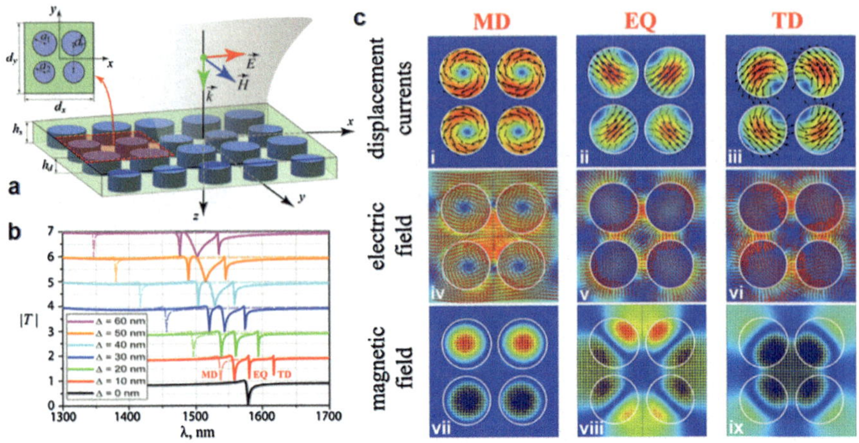

Fig. 5.6 Design and spectral response of the all-dielectric toroidal metasurface. **a** Schematic of the proposed metasurface. The geometries are: $d = 1250$ nm, $d_r = 540$ nm, $h_d = 120$ nm, $a = 240$ nm, and $a_1 = a_2 = a - \Delta$. **b** Transmission spectra of the all-dielectric metastructure for different degree of asymmetry (Δ). **c** Cross-cut profiles of displacement currents (black arrows), electric (red arrows), and magnetic (blue arrows) fields for the asymmetric cluster when $\Delta = 10$ nm. Adapted from (Tuz et al. 2018) with permission. Copyright ACS

all-dielectric metasurface consisting of periodically arranged asymmetric clusters of two pairs of dielectric nanodisks with different radii a_1 and a_2, in which this difference is specified as 'degree of asymmetry' $\Delta = |a_1 - a_2|$. Under x-polarized light illumination, from the set of transmission spectra presented in Fig. 5.6b, one can conclude that as the degree of asymmetry increases, all three pronounced resonances are still exist. Since the geometry of the asymmetric assembly is essentially anisotropic, both electric quadrupole (EQ) and toroidal dipole (TD) resonances appear only for the x-polarized excitation. Here, it should be noted that the toroidal response is technically related to the presence of trapped (dark) mode that emerges due to the symmetry breaking in the cluster. In practice, the trapped modes occur in the scattering and radiation analyses, and they exhibit free oscillations with a finite energy. Theoretically, the trapped mode is associated with an eigenvalue implanted in the continuous spectrum of the corresponding component (Evans et al. 1993). This combination of the continuous and discrete spectra leads to the rise of resonant properties with unique profiles (i.e., Fano, EIT, and PIT resonances). To investigate the behavior of the excited modes in the transmission spectra, one should calculate the induced current and field distribution maps for each mode. Figure 5.6c illustrates the numerically defined cross-cut patterns of displacement currents (black arrows), electric (red arrows), and magnetic (blue arrows) fields within the metamolecule at the corresponding resonant wavelengths. Obviously, when the metamolecule is symmetric ($\Delta = 0$), one can perceive a collective response of four longitudinal magnetic dipoles (MDs), where the displacement currents demonstrate a spinning behavior twisting around the center of each disk. By breaking the symmetry of the system ($\Delta \neq 0$), besides the MD, two extra resonances emerge in the transmission spectra of the metasurface (Fig. 5.6b). As can be seen in the displacement current distribution and the E- and H-field maps inside the all-dielectric assembly, the additional lineshapes are correlated with the EQ and TD modes. The latter can be recognized from a particular closed-loop arrangement of the displacement currents that penetrates all nanodisks in the cluster and surrounds the E-field lines circulating on a torus.

The need for the generation of substantially narrow resonant lineshapes has led researchers to develop novel concepts and use innovative methods (Sayanskiy et al. 2018; Kim et al. 2015; Tasolamprou et al. 2016; He et al. 2018). Among them, establishing a connection between the toroidal dipole resonance and the *bound states in the continuum (BIC)* in the context of all-dielectric metasurfaces enabled the excitation of high quality-factor (Q-factor) toroidal modes. In principle, the concept of *toroidal dipole BIC* is developed considering two eigenmodes of the all-dielectric metasurface that demonstrate an intrinsic toroidal dipole feature and an infinite lifetime. Such modes can be categorized as transverse and longitudinal toroidal dipole modes, in which in the context of BIC, such modes are associated with symmetry unprotected and protected BIC, respectively. The toroidal BIC supported by the symmetric metasurface has a potential to be turned into a toroidal dipole mode with ultranarrow spectral linewidth. This principle, indeed, confirms that such a sharp mode will be originated from toroidal dipole leaky resonances (Hsu et al. 2016). Figure 5.7a illustrates an example schematic rendering of the all-dielectric toroidal BIC metasurface. The quasi-infinite metastructure consists of a periodic array of silicon two-member

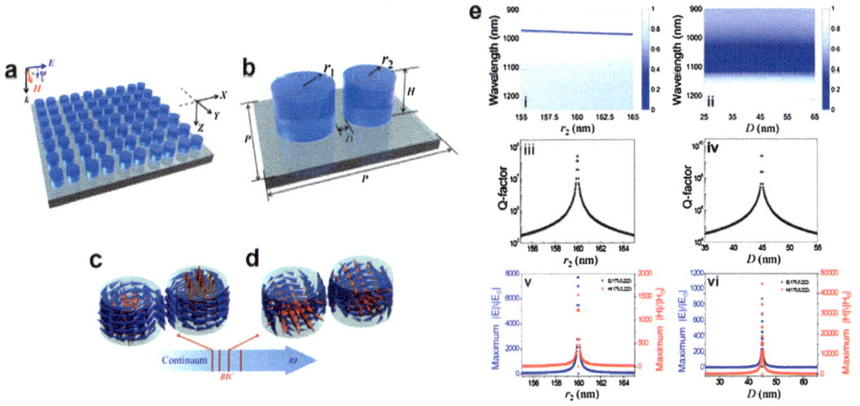

Fig. 5.7 The design and spectral response of a toroidal dipole BIC-resonant all-dielectric metasurface. **a** Schematic of the all-dielectric metasurface consisting of periodic arrays of silicon nanodisks. **b** A detailed rendering of the geometrical parameters of the all-dielectric unit cell. **c** and **d** Two toroidal dipole BIC eigenmodes of the photonic metasurface with their toroidal dipole moments along the y- and z-axes, respectively. **e** The dependence of transmissions on (i) r_2 and (ii) D. (iii) and (iv) The quantified Q-factors for the resonances around 1120 and 1213 nm, respectively. (v) and (vi) The E-field and H-field enhancements of both modes at a location 2 nm away from the surface of silicon dimer, where $r_1 = r_2 = 160$ nm, $P = 730$ nm, and $H = 220$ nm. Adapted from (He et al. 2018) with permission. Copyright APS

nanodisk dimers configured in a square lattice. Figure 5.7b displays the dimer structure with the corresponding geometrical parameters. The vectorial maps of E- and H-fields in Figs. 5.7c, d demonstrate a distinct toroidal feature within the unit cell of the metasurface, expressing themselves as poloidal currents and circular closed-loop magnetic fields. When the dimer is symmetric, the photonic metasurface efficiently supports toroidal dipole BIC moments. By breaking the symmetry of the metaplatform, the excited dipole turns into leaky resonances. For instance, by changing r_2, one can excite two resonances in the wavelength range of 900–1250 nm for an E-field in the x-direction (E_x) (Fig. 5.7e(i)). When $r_1 = r_2$, the linewidth of the toroidal dipole BIC resonance (around 1120 nm) becomes narrower and its Q-factor diverges, as shown Fig. 5.7e(iii). This behavior strongly verifies the formation of the toroidal dipole BIC. On the other hand, the other mode can be turned into a symmetry unprotected toroidal dipole BIC through modifying the distance D between the proximal nanodisks (Figs. 5.7e(ii) and (iv)). Importantly, the quantified Q-factors from the eigenmode analyses of the asymmetric metasurface confirm that two toroidal dipole BIC states can be turned into toroidal dipole leaky resonances when the symmetry of the nanodisks is broken. In addition, the excitation of such high Q-factor toroidal dipole BICs is associated with the robust near-field enhancements (Figs. 5.7e(v) and (vi)). The maximum E- and H-field enhancements indicate the potential of this approach for having strong light-matter interaction.

References

A. Ahmadivand, B. Gerislioglu, Directional toroidal dipoles driven by oblique poloidal and loop current flows in plasmonic meta-atoms. J. Phys. Chem. C **122**, 24304–24308 (2018)

A. Ahmadivand, B. Gerislioglu, N. Pala, Large-modulation-depth polarization-sensitive plasmonic toroidal terahertz metamaterial. IEEE Photon. Technol. Lett. **29**, 1860–1863 (2017)

A. Ahmadivand, B. Gerislioglu, A. Tomitaka, P. Manickam, A. Kaushik, S. Bhansali, M. Nair, N. Pala, Extreme sensitive metasensor for targeted biomarkers identification using colloidal nanoparticles-integrated plasmonic unit cells. Biomed. Opt. Express **9**, 373–386 (2018)

A.A. Basharin, M. Kafesaki, E.N. Economou, C.M. Soukoulis, V.A. Fedotov, V. Savinov, N.I. Zheludev, Dielectric metamaterials with toroidal dipolar response. Phys. Rev. X **5**, 011036 (2015)

A. Boltasseva, H.A. Atwater, Low-loss plasmonic metamaterials. Science **331**, 290–291 (2011)

C.R. de Galarreta, A.M. Alexeev, Y.-Y. Au, M. Lopez-Garcia, M. Klemm, M. Cryan, J. Bertolotti, C.D. Wright, Nonvolatile reconfigurable phase-change metadevices for beam steering in the near infrared. Adv. Funct. Mater. **28**, 1704993 (2018)

V.M. Dubovik, V.V. Tugushev, Toroid moments in electrodynamics and solid-state physics. Phys. Rep. **187**, 145–202 (1990)

D.V. Evans, C.M. Linton, F. Ursell, Trapped mode frequencies embedded in the continuous spectrum. J. Mech. Appl. Math. **46**, 253–274 (1993)

Y. Fan, Z. Wei, H. Li, H. Chen, Costas M. Soukoulis, Low-loss and high-Q planar metamaterial with toroidal moment. Phys. Rev. B **87**, 115417 (2013).

V.A. Fedotov, A.V. Rogacheva, V. Savinov, D.P. Tsai, N.I. Zheludev, Resonant transparency and non-trivial non-radiating excitations in toroidal metamaterials. Sci. Rep. **3**, 2967 (2013)

B. Gerislioglu, A. Ahmadivand, Functional Charge Transfer Plasmon Metadevices. Research **2020**, 9468692 (2020)

B. Gerislioglu, A. Ahmadivand, N. Pala, Functional quadrumer clusters for swithing between Fano and charge transfer plasmons. IEEE Photon. Technol. Lett. **29**, 2226–2229 (2017)
B. Gerislioglu, A. Ahmadivand, N. Pala, Tunable plasmonic toroidal terahertz metamodulator. Phys. Rev. B **97**, 161405 (2018)
B. Gerislioglu, G. Bakan, R. Ahuja, J. Adam, Y.K. Mishra, A. Ahmadivand, The role of $Ge_2Sb_2Te_5$ in enhancing the performance of functional plasmonic devices. Mater. Today Phys. **12**, 100178 (2020)
Y. He, G. Guo, T. Feng, Y. Xu, A.E. Miroshnichenko, Toroidal dipole bound states in the continuum. Phys. Rev. B **98**, 161112 (2018)
C.W. Hsu, B. Zhen, A.D. Stone, J.D. Joannopoulos, M. Soljacic, Bound states in the continuum. Nat. Rev. Mater. **1**, 16048 (2016)
L. Jing, Z. Wang, B. Zheng, H. Wang, Y. Yang, L. Shen, W. Yin, E. Li, H. Chen, Kirigami metamaterials for reconfigurable toroidal circular dichroism. NPG Asia Mater. **10**, 888–898 (2018)
T. Kaelberer, V.A. Fedotov, N. Papasimakis, D.P. Tsai, N.I. Zheludev, Toroidal dipolar response in a metamaterial. Science **330**, 1510–1512 (2010)
A.V. Kildishev, A. Boltasseva, V.M. Shalaev, Planar photonics with metasurfaces. Science **339**, 1232009 (2013)
S.-H. Kim, S.S. Oh, K.-J. Kim, J.-E. Kim, H.Y. Park, O. Hess, C.-S. Kee, Subwavelength localization and toroidal dipole moment of spoof surface plasmon polaritons. Phys. Rev. B **91**, 035116 (2015)
Z. Liu, S. Du, A. Cui, Z. Li, Y. Fan, S. Chen, W. Li, J. Li, C. Gu, High-quality-factor mid-infrared toroidal excitation in folded 3D metamaterials. Adv. Mater. **29**, 1606298 (2017)
B. Luk'yanchuk, N. I. Zheludev, S. A. Maier, N. J. Halas, P. Nordlander, H. Giessen, C. T. Chong, The Fano resonance in plasmonic nanostructures and metamaterials. Nat. Mater. **9**, 707–715 (2010).
K. Marinov, A.D. Boardman, V.A. Fedotov, N. Zheludev, Toroidal metamaterial. New J. Phys. **9**, 324 (2007)
B. Ögüt, N. Talebi, R. Vogelgesang, W. Sigle, P.A. Van Aken, Toroidal plasmonic eigenmodes in oligomer nanocavities for the visible. Nano Lett. **12**, 5239–5244 (2012)
N. Papasimakis, V.A. Fedotov, N.I. Zheludev, S.L. Prosvirnin, Metamaterial analog of electromagnetically induced transparency. Phys. Rev. Lett. **101**, 253903 (2008)
N. Papasimakis, V.A. Fedotov, K. Marinov, N.I. Zheludev, Gyrotropy of a metamolecule: wire on a torus. Phys. Rev. Lett. **103**, 093901 (2009)
J.B. Pendry, D. Schurig, D.R. Smith, Controlling electromagnetic fields. Science **312**, 1780–1782 (2006)
A. Sayanskiy, M. Danaeifar, P. Kapitanova, A.E. Miroshnichenko, All-Dielectric Metalattice with Enhanced Toroidal Dipole Response. Adv. Opt. Mater. **6**, 1800302 (2018)
A.J. Silenko, Electric current multipole moments in classical electrodynamics. Prog. Theor. Phys. **101**, 875–884 (1999)
C.M. Soukoulis, M. Wegener, Past achievements and future challenges in the development of three-dimensional photonic metamaterials. Nat. Photon. **5**, 523–530 (2011)
A.C. Tasolamprou, O. Tsilipakos, M. Kafesaki, C.M. Soukoulis, E.N. Economou, Toroidal eigenmodes in all-dielectric metamolecules. Phys. Rev. B **94**, 205433 (2016)
V.R. Tuz, V.V. Khardikov, Y.S. Kivshar, All-dielectric resonant metasurfaces with a strong toroidal response. ACS Photon. **5**, 1871–1876 (2018)
S. Xu, A. Sayanskiy, A.S. Kupriianov, V.R. Tuz, P. Kapitanova, H.-B. Sun, W. Han, Y.S. Kivshar, Experimental Observation of Toroidal Dipole Modes in All-Dielectric Metasurfaces. Adv. Opt. Mater. **7**, 1801166 (2019)
S. Yang, Z. Liu, L. Jin, W. Li, S. Zhang, J. Li, C. Gu, Surface plasmon polariton mediated multiple toroidal resonances in 3D folding metamaterials. ACS Photon. **4**, 2650–2658 (2017)
G. Zhang, C. Lan, R. Gao, Y. Wen, J. Zhou, Toroidal dipole resonances in all-dielectric oligomer metasurfaces. Adv. Theory Simulations **2**, 1900123 (2019)
N.I. Zheludev, Y.S. Kivshar, From metamaterials to metadevices. Nat. Mater. **11**, 917–924 (2012)

Chapter 6
Toroidal Metadevices

Abstract Toroidal excitations in well-engineered media have recently considered as a promising way that feature futuristic optical technologies through controlling radiative losses in both plasmonic and photonic systems. As mentioned in previous Chapters, within the past decade, there has been extensive research over the excitation principles of toroidal multipoles, specifically toroidal dipole, in flatland and 3D metasystems. In this Chapter, we focus on revolutionary devices that have been put in practice based on this notion, including infrared photodetectors, deep ultraviolet (DUV) beam sources, and immunobiosensors. Besides, we argue the vacuum Rabi oscillations through strong plexciton dynamics in this context.

6.1 Photodetection: Enhancing the Responsivity Performance

Plasmon excitations in subwavelength metallic structures can decay in the form of dynamic electron-hole pairs (EHPs) in a few nano- or microseconds, which are useful for photocurrent generation. This approach has extensively been exploited to develop next-generation nanoplasmonic light-sensing devices, with substantial quantum efficiency and responsivity, in various strategic areas; including but not limited to missile warning, target recognition, time-gated distance measurements, light harvesting, modern biotechnology, and low-power wavelength division multiplexing for short-distance optical communication (Chen et al. 2011; Chalabi et al. 2014; Brongersma et al. 2015; Dong et al. 2016; Gerislioglu and Ahmadivand 2019). Such an interest in plasmonic photodetection mechanisms (e.g., hot-electron- and free carrier absorption (FCA)-based) is due to the extreme light confinement capability of resonant metallic nanostructures (Li and Valentine 2017). In the active regime, the generation and sweeping of electrons and holes (through applying forward and reverse bias, respectively) provide significant photocurrent and high-photon yield towards high-responsivity and high-efficiency photodetection systems across a wide spectral range (Martyniuk et al. 2014; Chen et al. 2016; Kim et al. 2018; Moskovits 2015).

A. Ahmadivand et al., *Toroidal Metamaterials*, Engineering Materials,
https://doi.org/10.1007/978-3-030-58288-3_6

From the optical physics perspective, the interaction of incident electromagnetic wave with the metallic constituents at the subwavelength limit gives rise to the coherent oscillations of electrons at the *d*-band of noble metals, or plasmons (Barnes et al. 2003; Gerislioglu and Ahmadivand 2019; Green and Pillai 2012). The ever-increasing demand for reaching faster operation and higher responsivity manipulate the focus of researches on developing plasmon-enhanced optoelectronic devices (e.g., photodetectors and phototransistors), where plasmonics has promisingly modified the electrical and spectral characteristics of photodetection systems from UV to THz frequencies (Deng and Li 2014; Chen et al. 2015; Gerislioglu et al. 2019). So far, in particular, several approaches have been implemented to boost the quantum efficiency and performance of plasmonic photodetectors even further, such as quantum-dots or 2D materials (e.g., graphene and MoS_2)-mediated metallic platforms (Wang et al. 2015, 2017; Lee et al. 2009; Koppens et al. 2014; Chang et al. 2010; Tang et al. 2017) and combination of hot carrier generation and FCA concepts (Tanzid et al. 2018).

Beyond that, to maximize the performance, one can consider to induce dark (also known as "nonradiating") resonances using carefully structured platforms (Stockman 2010; Wu et al. 2011; Bullock et al. 2016). In the excited nonradiating resonance limit, the scattering cross-section is suppressed while the absorption cross-section is enhanced. In spite of generating large amount of hot carriers (for instance, at the metal-dielectric interface), this approach is still insufficient, in terms of field confinement towards high photon-yields and photocurrents, to be able to develop low-dimensional integrable plasmonic and photonic systems. Recently, to address these impediments, the utilization of ghost (or nonradiative) spectral features, as a new class of dark resonant moments (Ahmadivand and Gerislioglu 2018; Ahmadivand et al. 2018), has been introduced (Ahmadivand et al. 2019). The strong localization of electromagnetic field, enhanced absorption cross-section, and low emission rate are the features enabled by dynamic toroidal dipole that have been utilized for the formation of active hot electrons (through nonradiative decay) and photocurrent enhancement.

Figure 6.1 shows the implementation example of the narrowband IR toroidal photodetection mechanism. As indicated in Fig. 6.1a, the device consists of asymmetric gold standalone pixels (to enable the robust charge-current configuration), in which the capacitive gaps between the central and peripheral resonators are 25 nm, and the height of the resonators are set to 40 nm. Since doped silicon (Si) remarkably absorbs near-IR illumination (Spitzer and Fan 1957; Kane 1956; Gomez et al. 2013), the periodic arrays of the plasmonic meta-atoms are located on top of a doped Si substrate to further increase the photodetection performance of the proposed platform. Figure 6.1b verifies (both numerically and experimentally) the excitation of distinct toroidal dipole at 2850 nm under transverse electric (TE) polarized light. In addition to the toroidal mode, a magnetic dipole is induced at 2150 nm. In the inset of Fig. 6.1b, scanning electron microscopy (SEM) images of the fabricated array are presented. As it was expectable, in Fig. 6.1c, the normalized absorption cross-section reaches its maximum value around 2850 nm, both for *p*- and *n*-type Si substrates (in both doping regimes, the carrier concentration is set to 2×10^{19} cm^{-3}). Here, the projected toroidal metastructure with *p*-type Si substrate shows

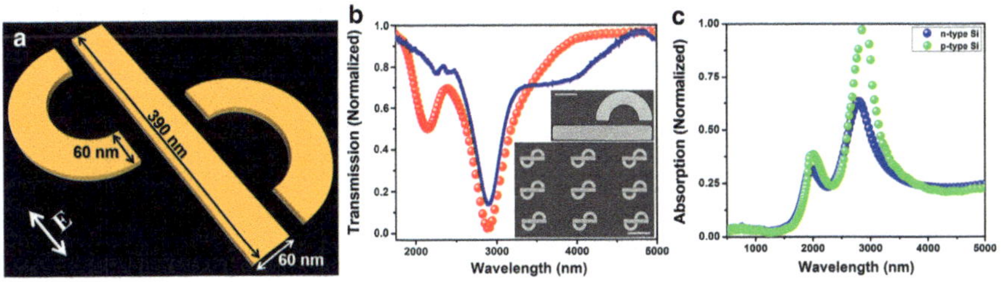

Fig. 6.1 Design and spectral response of the toroidal meta-device. **a** An artistic illustration of the asymmetric standalone structure, placed along the polarization of the impinging light. **b** Numerical (red) and experimental (blue) transmission spectra of the ordered unit-cells, indicating both toroidal and magnetic dipole modes at 2850 and 2150 nm, respectively. Insets: Zoom-out and zoom-in SEM images of the fabricated nanodevice, scale bars: 500 and 100 nm, respectively. **c** Normalized absorption cross-section of the toroidal meta-atom for both *p*- and *n*-type doped Si substrates. Adapted from (Ahmadivand et al. 2019) with permission. Copyright RSC

much higher absorption at the nonradiative resonance regime, owing to longer lifetime of the photoinduced carriers, excessive number of electrons at the Schottky interface (Chalabi et al. 2014; Gerislioglu et al. 2019), and increased carrier mobility of the Si substrate, leading to augmented photocurrent.

The formation of the toroidal dipole moment (as oblique charge-current arrangement) is visually illustrated in Fig. 6.2. As mentioned in Chap. 2, the mismatch between the flux directions of charges in the peripheral curve-shaped nanoantennas creates the head-to-tail spinning configuration of magnetic dipoles. Here, this claim is further corroborated with: (i) the induced surface current density across the meta-atom (Fig. 6.2a), (ii) the contribution of the induced toroidal dipole mode to the radiated power from plasmonic scatterer (Fig. 6.2b), and (iii) cross-sectional, vectorial charge-current map along the entire unit cell (Fig. 6.2c). Following, in light of these

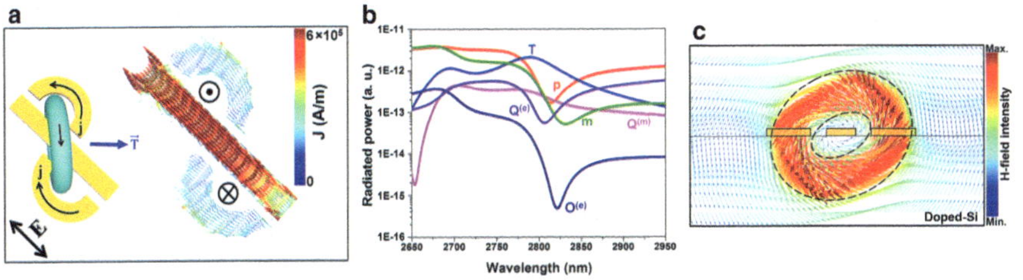

Fig. 6.2 Numerical verification of the induced toroidal dipole mode properties. **a** The vectorial displacement surface current density "J" across the unit cell under *y*-polarized beam. Inset: A schematic for the spinning charge-current configuration and the rotation direction of both J and induced toroidal feature ($\vec{T}$). **b** Radiated power of each individual EM multipoles generated with the meta-atom as a result of incidence. **c** Cross-sectional vectorial magnetic (H) field distribution to indicate the head-to-tail charge-current configuration. Adapted from (Ahmadivand et al. 2019) with permission. Copyright RSC

characteristics, the hot electron generation process within the considered toroidal plasmonic metadevice under near-IR illumination and applied bias is demonstrated. Theoretically, at the toroidal dipole position ($\lambda \sim 2850$ nm), the induced field squeezes into a tiny spot and enhances the field localization effect. Based on the hot electron generation principle (Chalabi et al. 2014; Brongersma et al. 2015), this provides a substantial augmentation in the photogenerated current. Indeed, when the energy ($h\nu$) of the photoinduced plasmons becomes sufficiently large, the decayed electrons will have enough energy to ascend over the innately formed Schottky barrier and transmitted into the doped Si layer where they get collected through the applied bias (0–500 mV).

In Fig. 6.3, the impact of the induced toroidal mode on the carrier generation was validated both numerically and experimentally. Specifically, the stimulation of intense and confined plasmons along the metallic nanostructures gives rise to the generation of the hot electrons and subsequently photocurrent enhancement. With the help of reduced electron-electron scattering within the metadevice, the number of hot electrons transferred to the doped Si is incremented and expedited before they recombined (Wu et al. 2015; Baker-Finch et al. 2014). On top of that, due to higher carrier mobility in *p*-type Si (mainly because of FCA) and longer lifetime of carriers

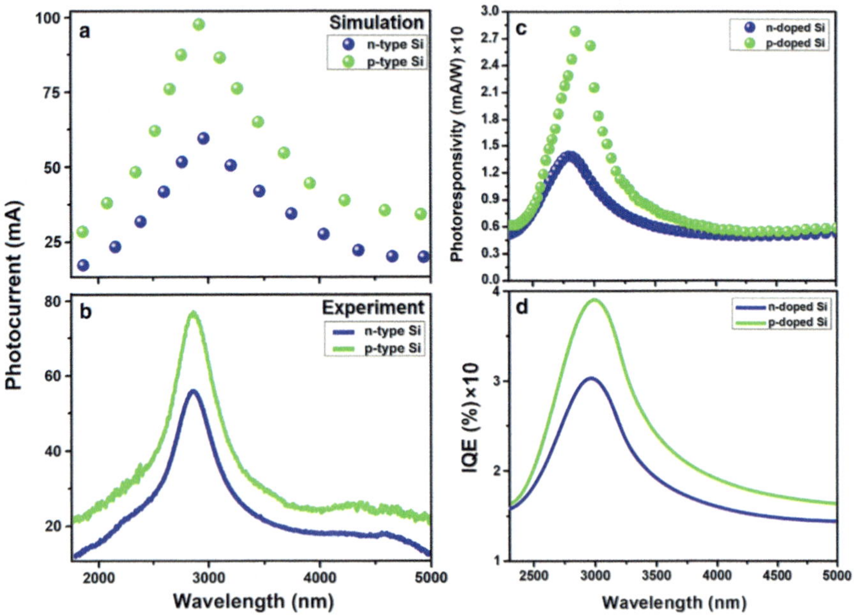

Fig. 6.3 Performance of the toroidal plasmonic photodetector. **a** Numerical and **b** experimental magnetoelectric currents at the electrodes for both *n*- and *p*-type Si substrates. Here, the drain-source voltage is fixed to ±5 mV. **c** Photoresponsivity and **d** IQE characteristics of the metadevice as a function of incidence for both substrates. Adapted from (Ahmadivand et al. 2019) with permission. Copyright RSC

(~100 μs), the proposed device produces much larger photocurrent in the boron-doped Si regime (see Figs. 6.3a, b). Moreover, in Fig. 6.3c, numerically obtained photoresponsivity of the toroidal system is demonstrated for both doping conditions. As expected, remarkable photoresponsivity values are quantified as ~14.5 mA W^{-1} and ~29 mA W^{-1} for the *n*- and *p*-type devices, respectively. Lastly, as one of the most important photodetection parameters, the internal quantum efficiency (IQE) of the metasystem is calculated using the following equation (Gerislioglu et al. 2019): $IQE = \left(I_p/q\right)/(S_{abs}/h\upsilon)$, where S_{abs} is the absorbed optical power which leads to the induced photocurrent, h is the Planck's constant, υ is the frequency of the incident electromagnetic wave, and q is the elementary charge. Basically, IQE can be defined as the ratio between the total number of charge carriers contributing to the photocurrent (I_p) and the total number of photons absorbed by the platform. The results showed that an IQE of 38.5% is achieved for the *p*-type toroidal metadevice, while this value is lessened to ~30% for the *n*-type photodetector (see Fig. 6.3d). Nevertheless, one can further increase the value of IQE of the proposed toroidal metadevice using atomically-thin monolayers (e.g., MXenes and graphene) (Echtermeyer et al. 2016; Chen et al. 2017; Velusamy et al. 2019).

The demonstrated toroidal photodetector was the first example of its kind. The numerical and experimental results explicitly demonstrated how the implemented technology can substantially boost the photocurrent generation and photon yield in a metasystem. All of these features make the devised instrument as a promising competitor for ever-growing plasmonic and nanophotonic IR beam sensing world.

6.2 Nonlinear Lasing: Deep Ultraviolet Source

The harmonic signal generation is one of the major application areas in nonlinear optics, which has received lots of attention recently (Smirnova and Kivshar 2016; Hooper et al. 2019; Carletti et al. 2019). This phenomena mainly based on light-matter interaction (e.g., extreme confinement of the incident electromagnetic field) to effectively convert multiple low-energy photons into a single high-energy photon (Boyd 2003). Researchers have shown that the nonlinear optical effects can be amplified through designing nanoplasmonic and nanophotonic building blocks (Kauranen and Zayats 2012; Koshelev et al. 2020), towards a wide range of applications, including but not limited to lasing (Melentiev et al. 2013; Paarmann et al. 2015), beam shaping (Keren-Zur et al. 2018; Ren et al. 2020), and enhanced nano(bio) imaging (Rodrigues et al. 2014; Fang et al. 2017). When the size of structures becomes less than the incident light, phase-matching between the input and output fields is no longer achievable to produce short wavelength beam, which is also valid for conventional nonlinear crystals where they reach their limits in terms of transparency. Thus, new concepts need to be developed to overcome this bottleneck, especially for intense second and third harmonic nonlinear signal generation at the nanoscale.

Among different techniques, electromagnetic radiation with wavelengths between 100–190 nm and 190–280 nm, known as vacuum UV (VUV) and DUV, is one of the research topics to revolutionize optics (Jahani and Jacob 2016; Arbabi et al. 2015; Sivis et al. 2013; Pfullmann et al. 2013). To date, many important scientific and industrial applications for radiation in these regimes have been demonstrated, including photodissociation, photochemistry, lithographic patterning, and spectroscopy (Leach et al. 2005; Wang and Liu 1998; Wu and Kumar 2007; Milazzo and Cecchetti 1969). Current high-gain, coherent high-energy UV beam sources can be listed as excimer lasers, prism-coupled devices, free-electron lasers, and synchrotrons (Ederer et al. 1975; Gortler and Strowitzki 2005; Chang et al. 2018). Additionally, supercontinuum generation, novel nonlinear crystals, and high harmonic generation in gases and solids are the other possible UV beam source candidates (Ermolov et al. 2015; Halasyamani and Zhang 2017; Ozawa et al. 2015). It is noteworthy to state that the practical use of the mentioned methods requires complex, large, and laboratory-scale equipment and expensive fabrication process, while they yield the generation of intense and coherent VUV and DUV beams. As discussed above, optical metasurfaces (or building blocks) can be considered as one of the promising solutions to address these shortcomings. For this purpose, several plasmonic and all-dielectric metasurfaces have been proposed and formed to realize UV light generation based on second and third harmonic signals (Melentiev et al. 2016; Makarov et al. 2016; Ahmadivand et al. 2019; Shibanuma et al. 2017). Pioneering studies showed that plasmonic or photonic UV light sources possess their own advantages and disadvantages on producing such short wavelengths. For instance, all-dielectric metasurfaces enable more efficient nonlinear signal generation due to their inherent loss-less optical response and low heating to the incident high power illumination; however they suffer from the intensity of the induced UV signal due to the thickness of the metasurface. On the other hand, plasmonic metasurfaces can provide extreme electromagnetic field confinement at subwavelength dimensions, nevertheless the efficiency of the nonlinear process is critically affected by intrinsically lossy nature of metals.

Lately, it has been demonstrated that novel mechanisms based on anapole, toroidal, and Fano resonances with unique advantages can be utilized to improve nonlinear signal generation capabilities of plasmonic metasurfaces, by confining the incident electromagnetic field in a tiny spot. Among them, toroidal meta-atoms have been proposed to produce DUV signal, nominally five times stronger than that of a conventional plasmonic dimer nanosystem (Ahmadivand et al. 2019). The proposed gold metasurface was fabricated on an indium tin oxide (ITO)/glass substrate, where the corresponding geometric parameters and cross-sectional schematic of the whole structure are given in Figs. 6.4a, b, respectively. Here, the underlying ITO layer is utilized because of possessing high third order susceptibility $\chi^{(3)}$ value, while the gold nanoresonators provide the required field enhancement for the harmonic signal generation. In Fig. 6.4c, the perspective view of the metasurface design is presented, with the direction of the induced toroidal dipole moment $\left(\vec{T}\right)$ and the formation of the charge-current configurations between the neighboring nanoresonators.

As indicated in Fig. 6.5a, the fundamental wavelength of the toroidal device is close to 785 nm, and both numerical and experimental transmission spectra show a close consistency. Next, to capture the physical origin of the induced resonance, a

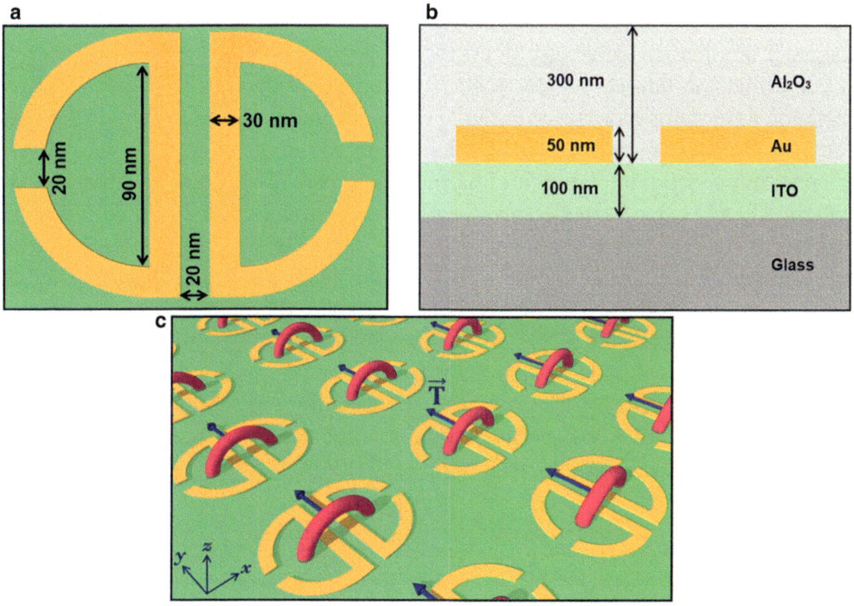

Fig. 6.4 Plasmonic toroidal metasurface for harmonic signal generation. **a** Top-view schematic of the meta-atom including geometric parameters. **b** Cross-sectional profile of the platform. **c** Perspective view of the metasurface showing the direction of the induced toroidal dipole mode ($\vec{T}$) and the formed charge-current configuration between the proximal resonators. Adapted from (Ahmadivand et al. 2019) with permission. Copyright ACS

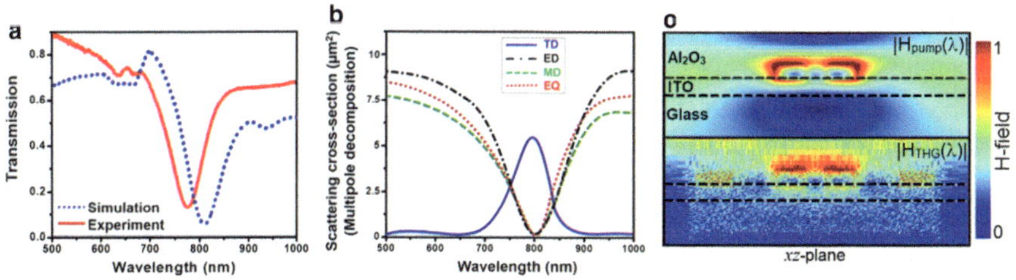

Fig. 6.5 Numerical simulation and device characterization. **a** Experimental (solid) and simulated (dotted) transmission spectrum of the toroidal metasurface. **b** Scattering cross-sections of different multipoles existed under illumination: toroidal dipole (TD), electric dipole (ED), magnetic dipole (MD), and electric quadrupole (EQ). **c** Cross-sectional, normalized magnetic field distribution at the fundamental ($|H_{pump}(\lambda)|$) and THG ($|H_{THG}(\lambda)|$) wavelengths. Adapted from (Ahmadivand et al. 2019) with permission. Copyright ACS

multipole decomposition analysis has been conducted (see Fig. 6.5b), in which the total scattering cross-section of the metasurface has contributions from a toroidal dipole (TD), an electric dipole (ED), a magnetic dipole (MD), and an electric quadrupole (EQ). Indeed, here, the dominant mode near the fundamental frequency can be ascribed as TD. As mentioned in Chap. 5, the discrepancy between the induced magnetic moments is necessary to create the desired charge-current loop configuration within the unit-cell. Clearly, the loop current induces a localized spinning magnetic field that penetrates into the ITO layer at both the fundamental (top) and third harmonic (bottom) wavelengths, demonstrated in Fig. 6.5c.

Next, the generation of the third harmonic signal at 262 nm was explored both numerically and experimentally, and the obtained results are presented in Figs. 6.6a, b, respectively. To provide a more realistic view for the third-order signal generation performance of the toroidal topology, it is compared with a conventional plasmonic system, which is a disk-dimer array. As anticipated, the obtained third harmonic signal generation from the dimer array is much weaker around a similar resonance position, in spite of exerting much higher average pump power. Consequently, the provided intensities in Fig. 6.6a were scaled according to the quantified effective third-order susceptibility values. These values were calculated from the experimental data using the following equation (Reintjes 1984; Ponomarenko 2012):

$$\chi_{eff}^{(3)} = \left(\frac{\varepsilon_0 c \lambda n_\omega w_0^2}{6 l P_\omega}\right)\left(\frac{n_\omega n_{3\omega} P_{3\omega}}{P_\omega}\right)^{1/2} \tag{6.1}$$

where ε_0 is the vacuum permittivity, c is the speed of light, λ is the fundamental wavelength, n_ω and $n_{3\omega}$ are the refractive indices at the fundamental ($\lambda = 785$ nm) and third harmonic ($\lambda_{THG} = 262$ nm) resonances, w_0 is the beam waist radius, P_ω and $P_{3\omega}$ are the peak powers at the fundamental and third harmonic, respectively, and l is the interaction length. Overall, the calculation showed that the effective third-order susceptibility of the toroidal metasurface ($\chi^{(3)}_{eff(toroid)}$) is $1.2 \times 10^{-21}\ m^2 \cdot V^{-2}$, which is 2.2- and 3.1-times greater than that of the disk-dimer array and the unpatterned ITO sublayer, respectively.

This particular study clearly verified the potential of the novel toroidal metallodielectric media towards leading nonlinear signal generation within the high energy

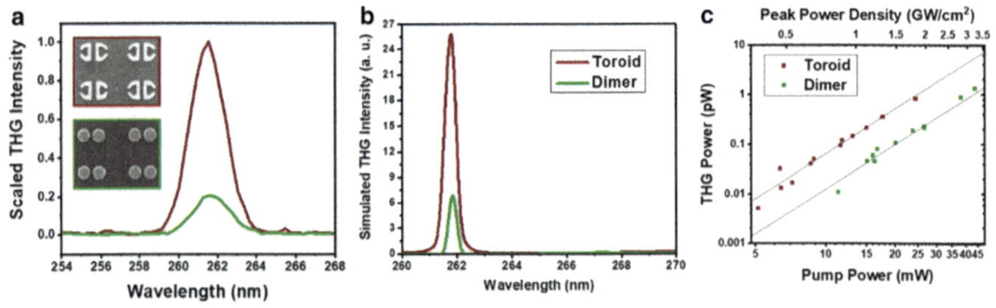

Fig. 6.6 Nonlinear characterization of the device. **a** Experimental and **b** simulated third harmonic spectrum of the toroidal metasurface (red) and dimer array (green). **c** Power dependence of the THG of the toroidal metasurface (red) and dimer array (green) with respect to the pump power in a log-log plot. Here, the lines are representing a third order power dependence. Adapted from (Ahmadivand et al. 2019) with permission. Copyright ACS

regions of the electromagnetic spectrum. Despite the nonradiative modes (e.g., Fano and anapole resonances) provide strong nonlinear enhancements, exciting them at much shorter wavelengths is quite challenging. However, the toroidal dipole, with its charge-current configuration feature, can be excited within the visible regime and induce robust electromagnetic field localization for harmonic signal generation in DUV region.

6.3 Immunosensors: Beyond Conventional Detection Limits

Plasmonic and photonic metachips have provided cost-effective, sensitive, and reliable label-free biomarker detection modalities by enabling robust confinement of electromagnetic field (Li et al. 2019; Cai et al. 2019; Leitis et al. 2019; Gerislioglu et al. 2020; Liu et al. 2010). These characteristics have facilitated the identification of a broad range of biomolecules (e.g., antibiotics, lipids, envelope proteins, viruses, DNA, enzymes) at ultralow densities with high precision. Although various types of metasensors (e.g., Fano resonant, electromagnetically-induced transparency (EIT) resonant etc.) have been developed to detect low-weight biological targets at reasonably low concentrations (Ruan et al. 2017; Ahmadivand 2019; Liu et al. 2010), these devices still do not meet the requirements for the identification of ultralow-weight proteins and viruses (e.g., MS2, PRD1, Zika, tau protein, beta-amyloid) at the early-stage of diseases, because of their weak limit of detection (LOD).

Recently, as an alternative to the conventional platforms, a new class of label-free, onsite, selective, and extremely sensitive biosensing technology is introduced based on toroidal electrodynamics (Ahmadivand et al. 2017, 2018, 2020; Gupta et al. 2017). From the theoretical physics perspective, due to its ultranarrow lineshape, pronounced quality factor (Q-factor), and strongly squeezed local field enhancement, the toroidal dipole provides much higher sensitivity to the perturbations in the dielectric permittivity of the surrounding medium, which have led to the rise of advanced biochemical and immunosensors with remarkable LOD and figure of merit (FOM). This can be better understood by considering the radiated electric field from conventional $\left\{E_p = n^2 k_0^2 \left(\hat{r} \times \hat{r} \times \frac{\vec{p}}{r}\right)\right\}$ and toroidal $\left\{E_T = n^3 k_0^3 \left(\hat{r} \times \hat{r} \times \frac{\vec{T}}{r}\right)\right\}$ scatterers, where $\hat{r}$ is the vector linking the location of the dipolar moment with the observer and n is the refractive index of the media. As indicated in equations above, slight variations in the dielectric permittivity of the media have a huge impact on the toroidal dipole limit, which allows the quantitative fingerprinting of extreme subwavelength multimolecular aggregates, and ultralow-weight biomolecules, biomarker proteins and antibodies (Ahmadivand et al. 2017, 2018, 2019, 2020; Gupta et al. 2017).

Among recently reported toroidal metasensors (Ahmadivand et al. 2017, 2018, 2019; Gupta et al. 2017), in this section, we highlight a promising study about the implementation of a plasmonic toroidal metasensor in the mid-infrared (MIR) regime. In this study (Ahmadivand et al. 2019), the researchers presented a precise and rapid way of detecting ultralow-weight Kantrex biomolecules antibiotic molecules

(~0.6 KDa) at attomolar densities, with the help of robustly squeezed electromagnetic fields within the designated toroidal plasmonic meta-atoms. MIR spectrum perfectly covers the vibrations of most of the biological substances, including but not limited to envelope proteins, organisms, enzymes, DNA, and lipids (Yu et al. 2006). Even if the vibrational signals of these biomolecules are inherently weak, MIR spectroscopy facilitates reaching their fingerprints as a non-invasive, non-poisonous, and non-destructive label-free method (Yu et al. 2006; Zhu et al. 2018).

In Fig. 6.7a, a 3D schematic of the toroidal unit cell in the presence of Kantrex biomolecules is depicted and the inset represents the molecular structure of kanamycin sulfate. The SEM image of the fabricated arrays is presented in Fig. 6.7b. The transmission spectra for both experimental measurements and numerical calculations under *y*-polarized beam illumination are demonstrated in Fig. 6.7c, proving the excitation of a toroidal dipole at $\lambda \sim 5250$ nm ($\omega \sim 1904.7$ cm^{-1}). Conversely, for the *x*-polarized beam excitation, the toroidal feature disappears owing to weak discrepancy between the circulating magnetic fields and surface current densities in adjacent resonators. Next, as a proof of principle, the toroidal metasensor is covered with the Kantrex antibiotic molecules at different concentrations to assess the sensing capability of the biosensor. For this purpose, the detection of the antibiotic molecules around the toroidal meta-atoms is conducted using the transmission difference between two different regimes (namely, in the absence and presence of biomolecules) as:

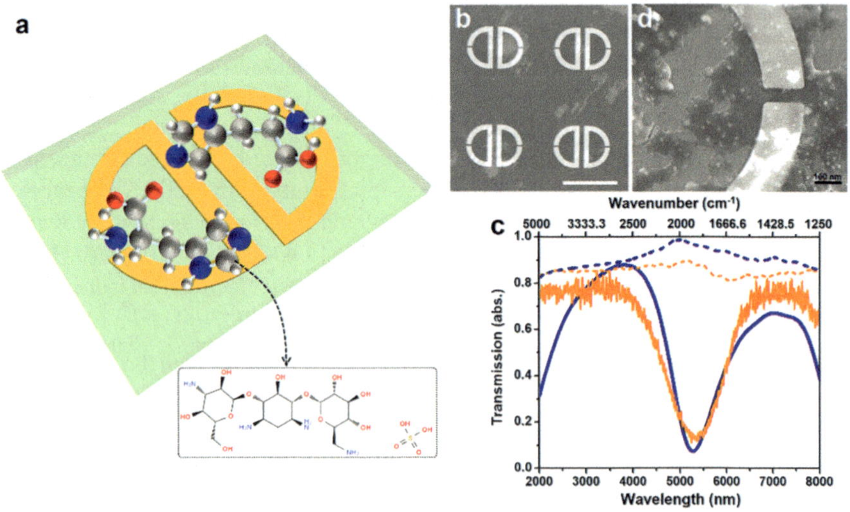

Fig. 6.7 Toroidal plasmonic MIR biosensor. **a** An artistic schematic of the toroidal meta-atom in the presence of kanamycin sulfate molecules. Here, the inset displays the molecular structure of the targeted molecules. **b** SEM image of the fabricated metasurface. The scale bar is 2 μm. **c** Numerical (blue) and experimental (orange) transmission spectra under y- (solid line) and x-polarized (dashed line) light. **d** The SEM image of an area of the fabricated structure in the presence of antibiotic molecules. Adapted from (Ahmadivand et al. 2019) with permission. Copyright APS

$$\Delta T(\omega) \equiv \left|t_{yy}^{water}(\omega)\right|^2 - \left|t_{yy}^{molecule}(\omega)\right|^2 \tag{6.2}$$

where t_{yy} is the tensor correlating the incident and transmitted electric fields through the metamaterial under y-polarized illumination. By introducing 10 μL of antibiotic solution to the plasmonic metamaterial specimens (see Fig. 6.7d), it is validated that the accumulation of Kantrex molecules at the capacitive gap regions dramatically changes the measured spectra.

More precisely, considering water as a reference, a prominent red-shift (in the range of 5400 nm < λ_t < 6000 nm (or 1515.1 cm^{-1} < ω_t < 1851.8 cm^{-1})) is observed in the spectral position of toroidal dipole, while the concentration of the targeted Kantrex molecules continuously increases from 0.1 to 10 fmol (Fig. 6.8a). Besides, in Fig. 6.8b, the displacement in the toroidal dipole position as a function of the concentration of Kantrex molecules is plotted, in which the slope of the red-shift of

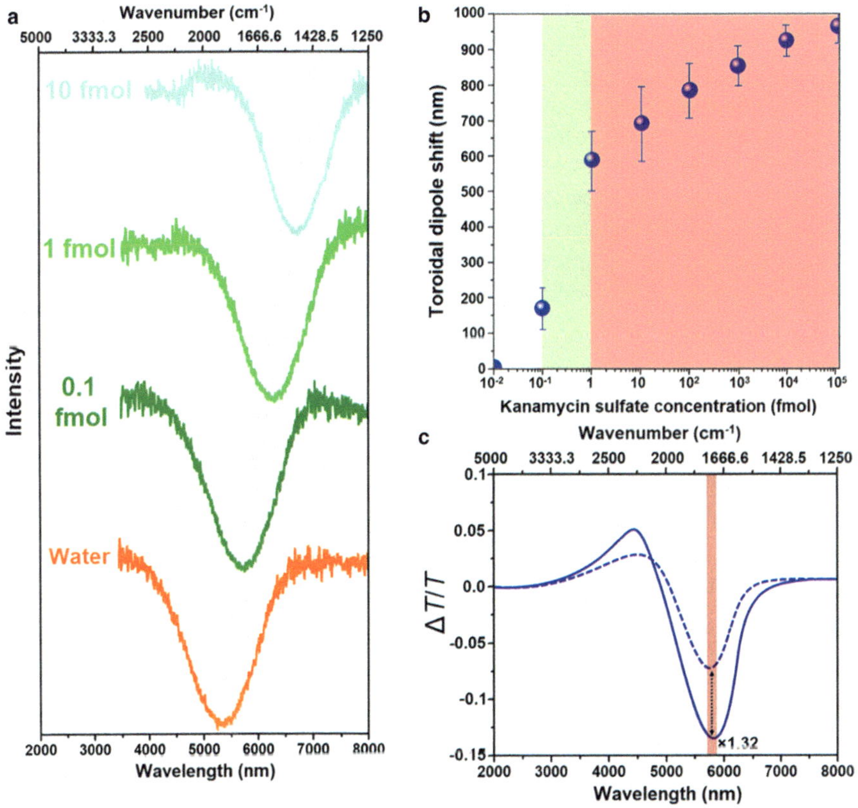

Fig. 6.8 Sensing characteristics of the metasensor. **a** Experimentally measured transmission spectra for different concentrations of the targeted molecules. **b** The induced toroidal dipole position and **c** the lifetime of the toroidal dipole as a function of the antibiotic concentration. Adapted from (Ahmadivand et al. 2019) with permission. Copyright APS

the asymmetric lineshape follows a relatively sharp inclination from 0.1 to 10 fmol (highlighted with green hue). This trend clearly verifies the LOD of the metasensor (~0.85 fmol (850 amol)). A moderate slope (shaded with red hue) was obtained for much denser concentrations of the molecules, because of their destructive impact on the formation of toroidal mode. Indeed, to explore the origin of the observed resonance shift, one can combine the permittivity variations and near-field coupling as (Joannopoulos 2008):

$$\frac{\Delta\lambda_T}{\lambda_T} = -\frac{1}{2}\left(\frac{\int_0^t \mathbf{E}_{\mathrm{T}}(\mathbf{r})(\hat{\varepsilon} - 1)\mathbf{E}_{\mathrm{T}}(\mathbf{r})d\mathbf{r}}{\int_0^\infty |\mathbf{E}_{\mathrm{T}}(\mathbf{r})|^2 d\mathbf{r}}\right) \tag{6.3}$$

where t is the thickness of the dropped liquid layer on the surface of the metachip, $\hat{\varepsilon}$ is the permittivity, and $|\mathbf{E}_{\mathrm{T}}(\mathbf{r})|$ is the near-field at the gaps. To model the Kantrex molecules around the system, the experimentally measured refractive index value ($n = 1.67$) was employed, and the approximate thickness (~5.8 nm) of the dispersed layer of molecules was defined based on the ellipsometric data. Besides, the extrapolated permittivity includes a nondispersive term (n^2), giving rise to an absorption peak (or transmission dip) along the MIR spectra. In Fig. 6.8c, the experimental results for before and after binding conditions are illustrated, where the alterations in the transmissivity ratios ($\Delta T/T$) of the toroidal moment are explicitly shown.

This work and other recent studies based on the use of toroidal metaplatforms for biosensing and immunosensing clearly demonstrate that how extremely localized field confinement in the unit cells leads to highly precise detection of low-weight molecules at extremely low-level of concentrations.

6.4 Plexciton Dynamics: Intensifying Ultrastrong Coupling

Strong coupling between plasmonic modes and excitonic states in subwavelength optical structures have received a growing interest towards the realization of pronounced plexciton dynamics. To understand the origin of this light-matter interaction, several approaches have been proposed, such as Fano interference (Artuso and Brytant 2008; Li et al. 2018), plasmon-enhanced emission and absorption (Achermann 2010; Antosiewicz et al. 2014), and hybridized platforms (Tanaka et al. 2010; Sugawara et al. 2006). Particularly, the tight confinement of incident light and the density of the conventional electromagnetic modes at subwavelength scales give rise to robust destructive and constructive interferences, which enables the formation of new spectral features (e.g., Fano resonance and vacuum Rabi oscillations) owing to the plexcitonic coupling (Bellessa et al. 2004; Gerislioglu and Ahmadivand 2020; Yan et al. 2017; Zhang et al. 2006; Zhao et al. 2016). Essentially, the vacuum Rabi oscillation is the most significant phenomenon that has been observed in such systems, in which the plasmonic systems are functionalized with multilevel excitonic materials, such as J- and H-aggregates (Wurtz et al. 2007; Spano 2015), dye molecules (Pockrand et al. 1982), quantum dots (QDs) (Akimov et al. 2007; Ahmadivand et al. 2019), and transition metal dichalcogenide monolayers (TMDCs) (Cuadra et al. 2018; Xue et al. 2016). Although there have been extensive research on understanding the fundamentals of plexciton dynamics and Rabi oscillations for integrated and low-threshold lasing tools, nonlinear harmonic signal generation, and quantum chemistry, finding much stronger coupling and the excitation of more prominent Rabi splitting is in a high demand, especially for multipoles with weaker far-field radiation patterns (e.g., anapole and toroidal multipoles).

To bridge this gap, as a pioneer work (Ahmadivand et al. 2019), researchers have introduced a NIR toroidal meta-atom (see Fig. 6.9a) to augment the plexcitonic coupling between the quantum emitters and the plasmons towards distinct Rabi splitting of the toroidal moment at room temperature. Here, lead (II) sulfide (PbS) QD aggregates with a bulk band gap of 0.41 eV and large exciton Bohr radius are utilized. As indicated by both experimental and numerical transmission spectra in Fig. 6.9b, the *p*-polarized plane wave illuminated platform (at normal incidence) shows a pronounced toroidal mode around ~0.95 eV, in which the SEM image of the fabricated unit cell is demonstrated in Fig. 6.9c.

Next, to explore the spectral response of the metasurface in the presence of organic QDs, the sensitivity of the toroidal moment to the coupling strength and localization intensity within the capacitive gaps is studied. Technically, to operate in the strong coupling regime, the coupling strength should exceed both the cavity loss rates and emitter scattering (Kleemann et al. 2017), which can be resolved by employing high *Q*-factor resonators. Besides, the coupling strength is proportional to the inverse-square of the mode volume ($V^{-1/2}$), thus, an extremely small cavity is necessary to localize the excitons emerging from organic molecules or 2D sheets (Zhao et al. 2010). In Fig. 6.9d, a rendering for the metasystem in the presence of commercially

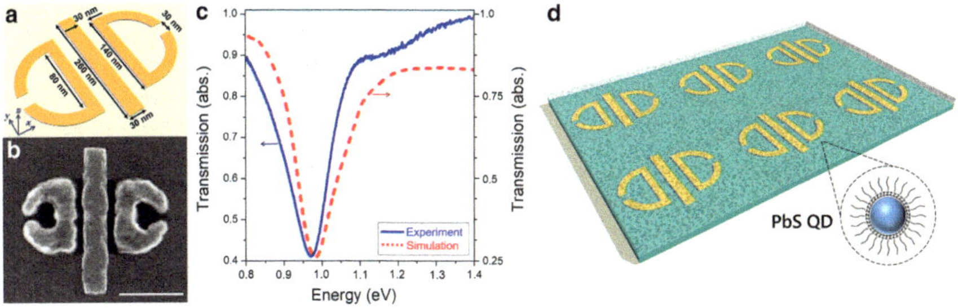

Fig. 6.9 Concept and the spectral response of the toroidal meta-atom. **a** A 3D schematic of the proposed meta-atom with the geometric parameters (not to scale). **b** Numerically calculated and experimentally obtained transmission spectra of the nanoplasmonic structure as a function of photon energy (in the absence of PbS QDs). **c** The SEM image of the fabricated toroidal unit-cell. The scale bar: 100 nm. **d** An artistic schematic of the proposed toroidal metasurface in the existence of PbS QDs. Adapted from (Ahmadivand et al. 2019) with permission. Copyright Wiley

available PbS QDs (with remarkable luminescence extreme in the NIR (Gates et al. 2005; Bakueva et al. 2004)) is presented.

By taking advantage of the strong absorption of nearly nonradiating toroidal moment, the Rabi splitting of the toroidal dipole feature (as a results of robust plexcitonic coupling) was investigated using both experimental and numerical techniques. As demonstrated in Fig. 6.10, the transmission spectra in the absence (i.e., vacuum (red solid line: experimental, red spheres: numerical)) and presence of QDs (i.e., PbS QDs (green solid line: experimental, green spheres: numerical)) indicate a substantial splitting ($\hbar\Omega$ ~ 150 meV) of the toroidal moment, due to existence of QDs and increased coupling rate between the plasmonic and excitonic states. It is noteworthy to mention that the above splitting value is calculated based on the following equation (Chikkaraddy et al. 2016):

$$\hbar\Omega = \sqrt{(4g^2 - [(\gamma_p - \gamma_e)/2]^2)} \tag{6.4}$$

where γ_p and γ_e are the dissipation rates of the uncoupled plasmons and excitons, respectively, and g is the coupling strength, defined as (Klem et al. 2005) $g = \mu_m\sqrt{4\pi\hbar Nc/\lambda_e\varepsilon\varepsilon_0 V}$, where $\mu_m = 17.5D$ is the transition dipole moment of PbS QDs (Hertzog et al. 2019), and N, λ_e, c, ε, V are the number of excitons, wavelength of excitons, velocity of light in vacuum, dielectric permittivity, and mode volume, respectively. In addition, the mode volume can be defined as $V = (\lambda/10n)^3$ (Han et al. 2018), where n is the refractive index of the media.

Furthermore, following the theoretical approach provided for the plasmonic effect on quantum coherence in QDs-mediated systems under dipole-dipole interaction (Shopova et al. 2011), the energy variations as a function detuning (Δ) are plotted in Fig. 6.10b. Based on the obtained dispersion profile, one can say that the plasmonic and excitonic extremes forms an anticrossing arc across the minimum detuning for

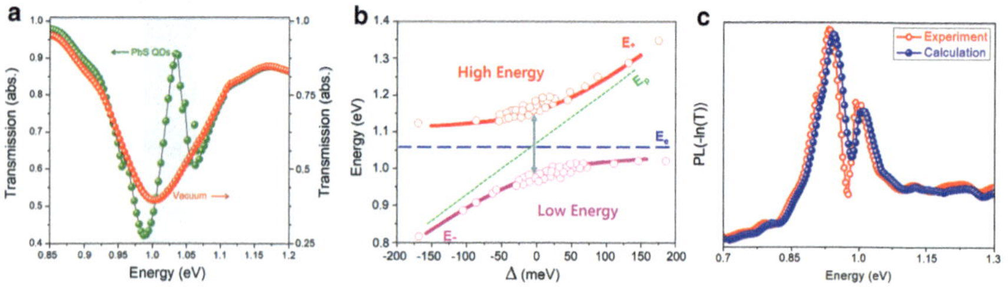

Fig. 6.10 Characteristics of the plexcitonic coupling in the toroidal meta-atom. **a** Calculated and measured transmission spectra of the toroidal metasurface in the absence and presence of QDs (with the average diameter of 6.4 nm). **b** Dispersion of plexciton with low and high energy bands as a function of detuning (Δ). Here, the dashed lines indicate the energy of the exciton resonance (E_e) of the QDs and plasmonic meta-atom (E_p). The gray arrow represents the Rabi splitting energy (~150 meV) of the hybrid toroidal platform. **c** Experimental and numerical photoluminescence (PL) spectra of the toroidal metasystem in the presence of QDs. Adapted from (Ahmadivand et al. 2019) with permission. Copyright Wiley

both high energy (E_+) and low energy (E_-) parts, validating the $\hbar\Omega \sim 150$ meV splitting of the toroidal dipole for the QDs with the average diameter of 6.4 nm. This trend can be elucidated by employing the robust plexcitonic coupling in the QDs-mediated plasmonic metadevice, in which the detuning of the toroidal dipole moment from the center of the exciton emission line provides a strong Purcell factor enhancement.

Beyond that the photoluminescence (PL) spectra of the toroidal metasystem in the existence of QDs with the average diameter of 6.4 nm was obtained through the probability of electron occupation in a given state using the estimated photonic density of states (PDOS). This probability calculation can be implemented by correlating transition matrix elements (Wang et al. 2019). In theory, the emission decay rate related to the PDOS of the photonic structure, which enables tailoring the internal dynamics of a quantum object-mediated system through photonic media that is resonant with radiative transition of the source. Based on this configuration, here, the toroidal metasurface creates an environment with a high Q-factor and the induced mode is confined within a small volume that boosts the corresponding PDOS, leading to the Purcell factor enhancement of luminescence as (Hechster and Sarusi 2017):$F_p = 3/4\pi^2(\lambda/n)^3 Q/V$, where λ is the wavelength, n is the refractive index of the surrounding environment, and V is the volume of the excited mode. As indicated in Fig. 6.10c, the experimental and numerical PL spectra for the QDs-functionalized toroidal plasmonic metasystem show a good agreement, which directly verifies the Purcell effect, plasmon-exciton interaction, and the quantified PDOS.

References

A. Ahmadivand, B. Gerislioglu, Directional toroidal dipoles driven by oblique poloidal current flows in plasmonic meta-atoms. arXiv preprint arXiv: 1807.07699 (2018)

A. Ahmadivand, R. Sinha, B. Gerislioglu, N. Pala, Sensor platform based on toroidal resonances for rapid detection of biomolecules. U.S. Patent 10288563 (2019)

A. Leitis, A. Tittl, M. Liu, B.H. Lee, M.B. Gu, Y.S. Kivshar, H. Altug, Angle-multiplexed all-dielectric metasurfaces for broadband molecular fingerprint retrieval. Sci. Adv. **5**, eaaw2871 (2019)

M. Achermann, Exciton-plasmon interactions in metal-semiconductor nanostructures. J. Phys. Chem. Lett. **1**, 2837 (2010)

A. Ahmadivand, B. Gerislioglu, Toroidal plasmonic meta-biosensors: Benchmarking against classical plasmonic biosensors and transducers. arXiv preprint arXiv: 1809.05602 (2018)

A. Ahmadivand, B. Gerislioglu, P. Manickam, A. Kaushik, S. Bhansali, M. Nair, N. Pala, Rapid detection of infectious envelope proteins by magnetoplasmonic toroidal metasensors. ACS Sens. **2**, 1359–1368 (2017)

A. Ahmadivand, B. Gerislioglu, A. Tomitaka, P. Manickam, A. Kaushik, S. Bhansali, M. Nair, N. Pala, Extreme sensitive metasensor for targeted biomarkers identification using colloidal nanoparticles-integrated plasmonic unit cells. Biomed. Opt. Express **9**, 373–386 (2018)

A. Ahmadivand, B. Gerislioglu, Z. Ramezani, Generation of magnetoelectric photocurrents using toroidal resonances: a new class of infrared plasmonic photodetectors. Nanoscale **11**, 13108–13116 (2019a)

A. Ahmadivand, M. Semmlinger, L. Dong, B. Gerislioglu, P. Nordlander, N.J. Halas, Toroidal dipole-enhanced third harmonic generation of deep ultraviolet light using plasmonic meta-atoms. Nano Lett. **19**, 605–611 (2019b)

A. Ahmadivand, B. Gerislioglu, Z. Ramezani, S.A. Ghoreishi, Attomolar detection of low-molecular weight antibiotics using midinfrared-resonant toroidal plasmonic metachip technology. Phys. Rev. Appl. **12**, 034018 (2019c)

A. Ahmadivand, B. Gerislioglu, Z. Ramezani, S.A. Ghoreishi, Demonstration of robust plexcitonic coupling in organic molecules-mediated toroidal meta-atoms. Adv. Optical Mater. **7**, 1901248 (2019d)

A. Ahmadivand, B. Gerislioglu, R. Ahuja, Y.K. Mishra, Terahertz plasmonics: the rise of toroidal metadevices towards immunobiosensing. Mater. Today **32**, 108–130 (2020)

A.V. Akimov, A. Mukherjee, C.L. Yu, D.E. Chang, A.S. Zibrov, P.R. Hemmer, H. Park, M.D. Lukin, Generation of single optical plasmons in metallic nanowires coupled to quantum dots. Nature **450**, 402–406 (2007)

T.J. Antosiewicz, S.P. Apell, T. Shegai, Plasmon-exciton interactions in a core-shell geometry: from enhanced absorption to strong coupling. ACS Photon. **1**, 454–463 (2014)

A. Arbabi, Y. Horie, M. Bagheri, A. Faraon, Dielectric metasurfaces for complete control of phase and polarization with subwavelength spatial resolution and high transmission. Nat. Nanotechnol. **10**, 937–943 (2015)

R.D. Artuso, G.W. Brytant, Optical response of strongly coupled quantum dot-metal nanoparticle systems: double peaked Fano structure and bistability. Nano Lett. **8**, 2106–2111 (2008)

S.C. Baker-Finch, K.R. McIntosh, D. Yan, K.C. Fong, T.C. Kho, Near-infrared free carrier absorption in heavily doped silicon. J. Appl. Phys. **116**, 063106 (2014)

L. Bakueva, I. Gorelikov, S. Musikhin, X.S. Zhao, E.H. Sargent, E. Kumacheva, PbS quantum dots with stable efficient luminescence in the near-IR spectral range. Adv. Mater. **16**, 926–929 (2004)

W.L. Barnes, A. Dereux, T.W. Ebbesen, Surface plasmon subwavelength optics. Nature **424**, 824–830 (2003)

J. Bellessa, C. Bonnand, J.C. Plenet, J. Mugnier, Strong coupling between surface plasmons and excitons in an organic semiconductor. Phys. Rev. Lett. **93**, 036404 (2004)

R. Boyd, *Nonlinear Optics* (Academic, Burlington, MA, 2003)

M.L. Brongersma, N.J. Halas, P. Nordlander, Plasmon-induced hot carrier science and technology. Nat. Nanotechnol. **10**, 25 (2015)

J. Bullock, M. Hettick, J. Geissbuhler, A.J. Ong, T. Allen, C.M. Sutter-Fella, T. Chen, H. Ota, E.W. Schaler, S. De Wolf, C. Ballif, Efficient silicon solar cells with dopant-free asymmetric heterocontacts. Nat. Energy **1**, 15031 (2016)

J. Cai, C. Zhang, C. Liang, S. Min, X. Cheng, W.-D. Li, Solution-processed large-area gold nanocheckerboard metasurfaces on flexible plastics for plasmonic biomolecular sensing. Adv. Opt. Mater. **7**, 1900516 (2019)

L. Carletti, S.S. Kruk, A.A. Bogdanov, C. De Angelis, Y.S. Kivshar, High-harmonic generation at the nanoscale boosted by bound states in the continuum. Phys. Rev. Res. **1**, 023016 (2019)

H. Chalabi, D. Schoen, M.L. Brongersma, Hot-electron photodetection with a plasmonic nanostripe antenna. Nano Lett. **14**, 1374–1380 (2014)

C.C. Chang, Y.D. Sharma, Y.-S. Kim, J.A. Bur, R.V. Shenoi, S. Krishna, D. Huang, S.-Y. Lin, A surface plasmon enhanced infrared photodetector based on InAs quantum dots. Nano Lett. **10**, 1704–1709 (2010)

Y. Chang, S. Yu, Q. Li, Y. Yu, H. Wang, S. Su, Z. Chen, L. Che, X. Wang, W. Zhang, D. Dai, G. Wu, K. Yuan, X. Yang, Tunable VUV photochemistry using vacuum ultraviolet free electron laser combined with H-atom Rydberg tagging time-of-flight spectroscopy. Rev. Sci. Instrum. **89**, 063113 (2018)

Q. Chen, D. Chitnis, K. Walls, T.D. Drysdale, S. Collins, D.R.S. Cumming, CMOS photodetectors integrated with plasmonic color filters. IEEE Photon. Technol. Lett. **24**, 197 (2011)

H. Chen, K. Liu, L. Hu, A.A. Al-Ghamdi, X. Fang, New concept ultraviolet photodetectors. Mater. Today **18**, 493–502 (2015)

H. Chen, H. Liu, Z. Zhang, K. Hu, X. Fang, Nanostructured photodetectors: from ultraviolet to terahertz. Adv. Mater. **28**, 403–433 (2016)

Z. Chen, X. Li, J. Wang, L. Tao, M. Long, S.-J. Liang, L.K. Ang, C. Shu, H.K. Tsang, J.-B. Xu, Synergistic effects of plasmonics and electron trapping in graphene short-wave infrared photodetectors with ultrahigh responsivity. ACS Nano **11**, 430–437 (2017)

R. Chikkaraddy, B. De Nijs, F. Benz, S.J. Barrow, O.A. Scherman, E. Rosta, A. Demetriadou, P. Fox, O. Hess, J.J. Baumberg, Single-molecule strong coupling at room temperature in plasmonic nanocavities. Nature **535**, 127–130 (2016)

J. Cuadra, D.G. Baranov, M. Wersall, R. Verre, T.J. Antosiewicz, T. Shegai, Observation of tunable charged exciton polaritons in hybrid monolayer WS2-plasmonic nanoantenna system. Nano Lett. **18**, 1777–1785 (2018)

K. Deng, L. Li, CdS nanoscale photodetectors. Adv. Mater. **26**, 2619–2635 (2014)

Y. Dong, Y. Gu, Y. Zou, J. Song, L. Xu, J. Li, J. Xue, X. Li, H. Zeng, Improving all-inorganic perovskite photodetectors by preferred orientation and plasmonic effect. Small **12**, 5622–5632 (2016)

T.J. Echtermeyer, S. Milana, U. Sassi, A. Eiden, M. Wu, E. Lidorikis, A.C. Ferrari, Surface plasmon polariton graphene photodetectors. Nano Lett. **16**, 8–20 (2016)

D.L. Ederer, E.B. Saloman, S.C. Ebner, R.P. Madden, The use of synchrotron radiation as an absolute source of VUV radiation. J. Res. Natl. Bur. Stand. Sect. A **79A**, 761–774 (1975)

A. Ermolov, K.F. Mak, M.H. Frosz, J.C. Travers, P.S.J. Russell, Supercontinuum generation in the vacuum ultraviolet through dispersive-wave and soliton-plasma interaction in a noble-gas-filled hollow-core photonic crystal fiber. Phys. Rev. A: At., Mol., Opt. Phys. **92**, 033821 (2015)

X. Fang, J. Wen, B. Bonello, J. Yin, D. Yu, Ultra-low and ultra-broad-band nonlinear acoustic metamaterials. Nat. Commun. **8**, 1288 (2017)

B.D. Gates, Q. Xu, M. Stewart, D. Ryan, C. Grant Willson, G.M. Whitesides, New approaches to nanofabrication: Molding, printing, and other techniques. Chem. Rev. **105**, 1171–1196 (2005)

B. Gerislioglu, A. Ahmadivand, The role of electron transfer in the nonlinear response of Ge2Sb2Te5-mediated plasmonic dimers. Photonics **6**, 52 (2019)

B. Gerislioglu, A. Ahmadivand, Theoretical study of photoluminescence spectroscopy of strong exciton-polariton coupling in dielectric nanodisks with anapole states. Mater. Today Chem. **16**, 100254 (2020)

B. Gerislioglu, A. Ahmadivand, J. Adam, Infrared plasmonic photodetectors: The emergence of high photon yield toroidal metadevices. Mater. Today Chem. **14**, 100206 (2019)

B. Gerislioglu, L. Dong, A. Ahmadivand, H. Hu, P. Nordlander, N.J. Halas, Monolithic metal dimer-on-film structure: new plasmonic properties introduced by the underlying metal. Nano Lett. **20**, 2087–2093 (2020)

D.E. Gomez, Z.Q. Teo, M. Altissimo, T.J. Davis, S. Earl, A. Roberts, The dark side of plasmonics. Nano Lett. **13**, 3722–3728 (2013)

A. Gortler, C. Strowitzki, Excimer lasers-the powerful light source in the UV and VUV. Laser Tech. J. **2**, 46–50 (2005)

M.A. Green, S. Pillai, Harnessing plasmonics for solar cells. Nat. Photon. **6**, 130–132 (2012)

M. Gupta, Y.K. Srivastava, M. Manjappa, R. Singh, Sensing with toroidal metamaterial. Appl. Phys. Lett. **110**, 121108 (2017)

P.S. Halasyamani, W. Zhang, Inorganic materials for UV and deep-UV nonlinear-optical applications. Inorg. Chem. **56**, 12077–12085 (2017)

X. Han, K. Wang, X. Xing, M. Wang, P. Lu, Rabi splitting in a plasmonic nanocavity coupled to a WS_2 monolayer at room temperature. ACS Photonics **5**, 3970–3976 (2018)

E. Hechster, G. Sarusi, Modeling the PbS quantum dots complex dielectric function by adjusting the Ek diagram critical points of bulk PbS. J. Appl. Phys. **122**, 024302 (2017)

M. Hertzog, M. Wang, J. Mony, K. Börjesson, Strong light-matter interactions: a new direction within chemistry. Chem. Soc. Rev. **48**, 937–961 (2019)

D.C. Hooper, C. Kuppe, D. Wang, W. Wang, J. Guan, T.W. Odom, V.K. Valev, Second harmonic spectroscopy of surface lattice resonances. Nano Lett. **19**, 165–172 (2019)

J. Yan, C. Ma, P. Liu, C. Wang, G. Yang, Generating scattering dark states through the Fano interference between excitons and an individual silicon nanogroove. Light: Sci. Appl. **6**, e16197 (2017)

S. Jahani, Z. Jacob, All-dielectric metamaterials. Nat. Nanotechnol. **11**, 23–36 (2016)

J.D. Joannopoulos, S.G. Johnson, J.N. Winn, R.D. Meade, *Photonic Crystals: Molding the Flow of Light* (Princeton Univ. Press, 2008)

J.F. Reintjes, *Nonlinear Optical Parametric Processes in Liquids and Gases* (1st ed. Academic Press, 1984)

E.O. Kane, Energy band structure in p-type germanium and silicon. J. Phys. Chem. Solids **1**, 82–99 (1956)

M. Kauranen, A.V. Zayats, Nonlinear plasmonics. Nat. Photon **6**, 737–748 (2012)

S. Keren-Zur, L. Michaeli, H. Suchowski, T. Ellenbogen, shaping light with nonlinear metasurfaces. Adv. Opt. Photon. **10**, 309–353 (2018)

D.H. Kim, W. Lee, J.-M. Myoung, Flexible multi-wavelength photodetector based on porous silicon nanowires. Nanoscale **10**, 17705–17711 (2018)

M.-E. Kleemann, R. Chikkaraddy, E.M. Alexeev, D. Kos, C. Carnegie, W. Deacon, A.C. de Pury, C. Große, B. de Nijs, J. Mertens, A.I. Tartakovskii, Strong-coupling of Wse2 in ultra-compact plamonic nanocavities at room temperature. Nat. Commun. **8**, 1296 (2017)

E.J.D. Klem, L. Levina, E.H. Sargent, PbS quantum dot electroabsrotpion modulation across the extended communications band 1200–1700 nm. Appl. Phys. Lett. **87**, 053101 (2005)

F.H.L. Koppens, T. Mueller, P. Avouris, A.C. Ferrari, M.S. Vitiello, M. Polini, Photodetectors based on graphene, other two-dimensional materials and hybrid systems. Nat. Nanotechnol. **9**, 780–793 (2014)

K. Koshelev, S. Kruk, E. Melik-Gaykazyan, J.-H. Choi, A. Bogdanov, H.-G. Park, Y. Kivhsar, Subwavelength dielectric resonators for nonlinear nanophotonics. Science **367**, 288–292 (2020)

S. Leach, H.-W. Jochims, H. Baumgartel, VUV photodissociation of ammonia: a dispersed fluorescence excitation spectral study. Phys. Chem. Chem. Phys. **7**, 900–911 (2005)

S.C. Lee, S. Krishna, S.R.J. Brueck, Quantum dot infrared photodetector enhanced by surface plasma wave excitation. Opt. Express **17**, 23160–23168 (2009)

W. Li, J.G. Valentine, Harvesting the loss: surface plasmon-based hot electron photodetection. Nanophotonics **6**, 177 (2017)

X. Li, L. Zhou, Z. Hao, Q.-Q. Wang, Plasmon-exciton coupling in complex systems. Adv. Opt. Mater. **6**, 1800275 (2018)

Z. Li, Y. Zhu, Y. Hao, M. Gao, M. Lu, A. Stein, A.-H.A. Park, J.C. Hone, Q. Lin, N. Yu, Hybrid metasurface-based mid-infrared biosensor for simultaneous quantification and identification of monolayer protein. ACS Photonics **6**, 501–509 (2019)

N. Liu, M. Mesch, T. Weiss, M. Hentschel, H. Giessen, Infrared perfect absorber and its application as plasmonic sensor. Nano Lett. **10**, 2342–2348 (2010a)

N. Liu, T. Weiss, M. Mesch, L. Langguth, U. Eigenthaler, M. Hirscher, C. Sonnichsen, H. Giessen, Planar metamaterial analogue of electromagnetically induced transparency for plasmonic sensing. Nano Lett. **10**, 1103–1107 (2010b)

S.V. Makarov, A.N. Tsypkin, T.A. Voytova, V.A. Milichko, I.S. Mukhin, A.V. Yulin, S.E. Putilin, M.A. Baranov, A.E. Krasnok, I.A. Morozov, P.A. Belov, Self-adjusted all-dielectric metasurfaces for deep ultraviolet femtosecond pulse generation. Nanoscale **8**, 17809–17814 (2016)

P. Martyniuk, J. Antoszewski, M. Martyniuk, L. Faraone, A. Rogalski, New concepts in infrared photodetector designs. Appl. Phys. Rev. **1**, 041102 (2014)

P.N. Melentiev, A.E. Afanasiev, A.A. Kuzin, A.S. Baturin, V.I. Balykin, Giant optical nonlinearity of a single plasmonic nanostructure. Opt. Express **21**, 13896–13905 (2013)

P.N. Melentiev, A.E. Afanasiev, A.A. Kuzin, V.M. Gusev, O.N. Kompanets, R.O. Esenaliev, V.I. Balykin, Split hole resonator: a nanoscale UV light source. Nano Lett. **16**, 1138–1142 (2016)

G. Milazzo, G. Cecchetti, Vacuum ultraviolet spectroscopy. Appl. Spectrosc. **23**, 197–203 (1969)

M. Moskovits, The case for plasmon-derived hot carrier devices. Nat. Nanotechnol. **10**, 6–8 (2015)

A. Ozawa, Z. Zhao, M. Kuwata-Gonokami, Y. Kobayashi, High average power coherent vuv generation at 10 MHz repetition frequency by intracavity high harmonic generation. Opt. Express **23**, 15107–15118 (2015)

A. Paarmann, I. Razdolski, A. Melnikov, S. Gewinner, W. Schollkopf, M. Wolf, Second harmonic generation spectroscopy in the reststrahl band of SiC using an infrared free-electron laser. Appl. Phys. Lett. **107**, 081101 (2015)

N. Pfullmann, C. Waltermann, M. Noack, S. Rausch, T. Nagy, C. Reinhardt, M. Kovacev, V. Knittel, R. Bratschitsch, D. Akemeier, A. Hutten, A. Leitenstorfer, U. Morgner, Bow-tie nano-antenna assisted generation of extreme ultraviolet radiation. New J. Phys. **15**, 093027 (2013)

I. Pockrand, A. Brillante, D. Mobius, Exciton-surface plasmon coupling: an experimental investigation. J. Chem. Phys. **77**, 6289 (1982)

M. Ren, W. Cai, J. Xu, Tailorable dynamics in nonlinear optical metasurfaces. Adv. Mater. **32**, 1806317 (2020)

S.P. Rodrigues, S. Lan, L. Kang, Y. Cui, W. Cai, Nonlinear imaging and spectroscopy of chiral metamaterials. Adv. Mater. **26**, 6157–6162 (2014)

B. Ruan, J. Guo, L. Wu, J. Zhu, Q. You, X. Dai, Y. Xiang, Ultrasensitive terahertz biosensors based on Fano resonance of a graphene/waveguide hybrid structure. Sensors **17**, 1924 (2017)

S.A. Ponomarenko, Fundamentals of Nonlinear Optics (Dalhousie University, 2012)

T. Shibanuma, G. Grinblat, P. Albella, S.A. Maier, Efficient third harmonic generation from metal-dielectric hybrid nanoantennas. Nano Lett. **17**, 2647–2651 (2017)

S.I. Shopova, R. Rajmangal, S. Holler, S. Arnold, Plasmonic enhancement of a whispering-gallery-mode biosensor for single nanoparticle detection. Appl. Phys. Lett. **98**, 243104 (2011)

M. Sivis, M. Duwe, B. Abel, C. Ropers, Extreme-ultraviolet light generation in plasmonic nanostructures. Nat. Phys. **9**, 304–309 (2013)

D. Smirnova, Y.S. Kivshar, Multipolar nonlinear nanophotonics. Optica **3**, 1241–1255 (2016)

F.C. Spano, Optical microcavities enhance the exciton coherence length and eliminate vibronic coupling in J-aggregates. The J. Chem. Phys. **142**, 184707 (2015)

W. Spitzer, H.Y. Fan, Infrared absorption in n-type silicon. Phys. Rev. **108**, 268 (1957)

M.I. Stockman, Nanoscience: dark-hot resonances. Nature **467**, 541–542 (2010)

Y. Sugawara, T.A. Kelf, J.J. Baumberg, M.E. Abdelsalam, P.N. Bartlett, Strong coupling between localized plasmons and organic excitons in metal nanovoids. Phys. Rev. Lett. **97**, 266808 (2006)

K. Tanaka, E. Plum, J.Y. Ou, T. Uchino, N.I. Zheludev, Multifold enhancement of quantum dot luminescence in plasmonic metamaterials. Phys. Rev. Lett. **105**, 227403 (2010)

X. Tang, G.F. Wu, K.W.C. Lai, Plasmon resonance enhanced colloidal HgSe quantum dot filterless narrowband photodetectors for mid-wave infrared. J. Mater. Chem. C **5**, 362–369 (2017)

M. Tanzid, A. Ahmadivand, R. Zhang, B. Cerjan, A. Sobhani, S. Yazdi, P. Nordlander, N.J. Halas, Combining plasmonic hot carrier generation with free carrier absorption for high-performance near-infrared silicon-based photodetection. ACS Photon. **5**, 3472–3477 (2018)

B. Gerislioglu, A. Ahmadivand, Towards infrared toroidal photodevices: a review. arXiv preprint arXiv: 1908.04804 (2019)

D.B. Velusamy, J.K. El-Demellawi, A.M. El-Zohry, A. Giugni, S. Lopatin, M.N. Hedhili, A.E. Mansour, E. Di Fabrizio, O.F. Mohammed, H.N. Alshareef, MXenes for plasmonic photodetection. Adv. Mater. **31**, 1807658 (2019)

J.-H. Wang, K. Liu, VUV photochemistry of CH4 and isotopomers. I. Dynamics and dissociation pathway of the H/D-atom elimination channel. J. Chem. Phys. **109**, 7105–7112 (1998)

W. Wang, A. Klots, D. Prasai, Y. Yang, K.I. Bolotin, J. Valentine, Hot electron-based near-infrared photodetection using bilayer MoS_2. Nano Lett. **15**, 7440–7444 (2015)

J. Wang, H. Fang, X. Wang, X. Chen, W. Lu, W. Hu, Recent progress on localized field enhanced two-dimensional material photodetectors from ultraviolet-visible to infrared. Small **13**, 1700894 (2017)

S. Wang, Q. Le-Van, F. Vaianella, B. Maes, S.E. Barker, R.H. Godiksen, A.G. Curto, J.G. Rivas, Limits to strong coupling of excitons in multilayer WS2 with collective plasmonic resonances. ACS Photonics **6**, 286–293 (2019)

B. Wu, A. Kumar, Extreme ultraviolet lithography: a review. J. Vac. Sci. Technol. B: Microelectron. Nanometer Struct. Proc. Meas. Phenom. **25**, 1743–1761 (2007)
C. Wu, A.B. Khanikaev, G. Shvets, Broadband slow light metamaterial based on a double-continuum Fano resonance. Phys. Rev. Lett. **106**, 107403 (2011)
K. Wu, J. Chen, J.R. McBride, T. Lian, Efficient hot-electron transfer by a plasmon-induced interfacial charge-transfer transition. Science **349**, 632–635 (2015)
G.A. Wurtz, P.R. Evans, W. Hendren, R. Atkinson, W. Dickson, R.J. Pollard, A.V. Zayats, W. Harrison, C. Bower, Molecular plasmonics with tunable exciton-plasmon coupling strength in J-aggregate hybridizes Au nanorod assemblies. Nano Lett. **7**, 1297–1303 (2007)
F. Xue, L. Chen, L. Wang, Y. Pang, J. Chen, C. Zhang, Z.L. Wang, MoS2 tribotronic transistor for smart tactile switch. Adv. Funct. Mater. **26**, 2104–2109 (2016)
C. Yu, A. Ganjoo, H. Jain, C.G. Pantano, J. Irudayaraj, Mid-IR biosensor: detection and finger-printing of pathogens on gold island functionalized chalcogenide films. Anal. Chem. **78**, 2500–2506 (2006)
W. Zhang, A.O. Govorov, G.W. Bryant, Semiconductor-metal nanoparticle molecules: hybrid excitons and the nonlinear Fano effect. Phys. Rev. Lett. **97**, 146804 (2006)
N. Zhao, T.P. Osedach, L.-Y. Chang, S.M. Geyer, D. Wanger, M.T. Binda, A.C. Arango, M.G. Bawendi, V. Bulovic, Colloidal PbS quantum dot solar cells with high fill factor. ACS Nano **4**, 3743–3752 (2010)
W. Zhao, S. Wang, B. Liu, I. Verzhbitskiy, S. Li, F. Giustiniano, D. Kozawa, K.P. Loh, K. Matsuda, K. Okamoto, R.F. Oulton, Exciton-plasmon coupling and electromagnetically induced transparency in monolayer semiconductors hybridized with Ag nanoparticles. Adv. Mater. **28**, 2709 (2016)
Y. Zhu, Z. Li, Z. Hao, C. DiMarco, P. Maturavongsadit, Y. Hao, M. Lu, A. Stein, Q. Wang, J. Hone, N. Yu, Q. Lin, Optical conductivity-based ultrasensitive mid-infrared biosensing on a hybrid metasurface. Light Sci. Appl. **7**, 67 (2018)

Printed by Printforce, the Netherlands